HARCOURT SCHOOL PUBLISHERS

Think Math!

Texas Edition

Student Work Text

Lesson Activity Book

Developed by Education Development Center, Inc. through National Science Foundation
Grant No. ESI-0099093

SCHOOL PUBLISHERS

Visit *The Learning Site!*
www.harcourtschool.com/thinkmath

This program was funded in part through the National Science Foundation under Grant No. ESI-0099093. Any opinions, findings, and conclusions or recommendations expressed in this program are those of the authors and do not necessarily reflect the views of the National Science Foundation.

Principal Investigator

E. Paul Goldenberg

Curriculum Design and Pedagogy Oversight

E. Paul Goldenberg | Lynn Goldsmith | Nina Shteingold

Research

Director: Lynn Goldsmith | Sabita Chopra | Suenita Lawrence
Nina Arshavsky | Sophia Cohen | Katherine Schwinden
Cynthia Char | Andrea Humez | Eugenia Steingold

Editorial

Director: Frances Fanning | Nicholas Bozard | Eric Karnowski

Writing

Director: Eric Karnowski

Jean Benson | Stacy Grossman | Paisley Rossetti
Abigail Branch | Andrea Humez | Nina Shteingold
Sara Cremer | Suenita Lawrence | Kate Snow
E. Paul Goldenberg | Debora Rosenfeld | Julie Zeringue

Graphics and Design

Directors: Laura Koval | Jessica Cummings | E. Charles Snow
and Korynn Kirchwey | Jennifer Putnam | Jenny Wong

Project Management

Directors: Eric Karnowski | Amy Borowko | Alexander Kirchwey | Kimberly Newson
and Glenn Natali | Nannette Feurzeig | Helen Lebowitz | David O'Neil
Kim Foster | June Mark | Cynthia Plouff

Mathematics Reviewers

Richard Askey, Professor of Mathematics, Emeritus
University of Wisconsin, Madison, Wisconsin

Harvey Keynes, Professor of Mathematics
University of Minnesota, Minneapolis, Minnesota

Roger Howe, Professor of Mathematics
Yale University, New Haven, Connecticut

David Singer, Professor of Mathematics
Case Western Reserve University, Cleveland, Ohio

Sherman Stein, Professor of Mathematics, Emeritus
University of California at Davis, Davis, California

Additional Mathematics Resource

Al Cuoco, Center Director, Center for Mathematics Education, Education Development Center, Newton, Massachusetts

Advisors

Peter Braunfeld | June Mark
David Carraher | Ricardo Nemirovsky
Carole Greenes | James Newton
Claire Groden | Judith Roitman
Deborah Schifter

Evaluators

Douglas H. Clements | Mark Jenness
Cynthia Halderson | Julie Sarama

iii

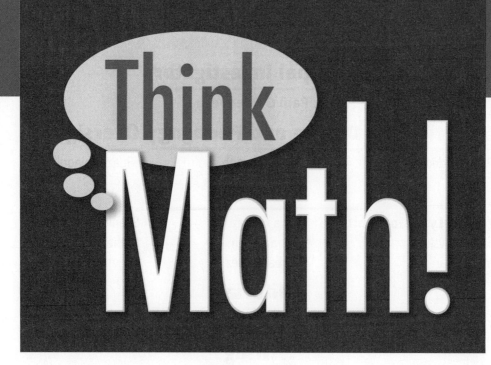

Chapter 1   Counting Strategies

Chapter 2   Working with 10

Think Math! Contents

Chapter 3   Place Value

Chapter 4   Addition and Subtraction with Place Value

Chapter 5　Probability and Data

Chapter 6　Measuring Time

Think Math! *Contents*

Chapter 7 Doubling, Halving, and Fractions

Chapter 8 Building Addition and Subtraction Fluency

Chapter 9   Two-Dimensional Figures and Spatial Sense

Chapter 10   Adding and Subtracting Larger Numbers

Think Math! *Contents*

Chapter 13   Exploring Multiplication and Division

Chapter 14   Comparing and Contrasting Three-Dimensional Figures

Think Math! *Contents*

Chapter 15 Capacity, Weight/Mass, and Temperature

Chapter 16 Multiplying and Dividing

Name _____

Counting Strategies
Paper Clip Patterns
 TEKS 2.6C

Make patterns with paper clips.

STEP 1 Creating Patterns

How many different patterns did you make? _____

Draw one of your patterns here.

_____

STEP 2 Describing Patterns

Use words to tell about the pattern you drew.

_____

Which is your favorite pattern? Tell about it. _____

_____

STEP 3 Extending Patterns

Continue one of your patterns. How did you know what to do?

_____

_____

_____

Investigation

Dear Family,

Today we started Chapter 1 of *Think Math!* In this chapter, I will explore numbers, number lines, patterns, skip-counting, addition, subtraction, and even multiplication. There are NOTES on the Lesson Activity Book pages to explain what I am learning every day.

Here are some activities for us to do together at home. These activities will help me understand numbers and counting patterns.

Love,

_____

Family Fun

What's My Number?

Work with your child to play a game called *What's My Number?* Your child will play this game later in this chapter.

- Tell your child you are thinking of a number from 1 to 8.

- Your child asks up to four *yes/no* questions to find the secret number. Each question should get rid of several numbers at once. Some good questions to ask are: "Is your number odd?" or "Is your number less than 5?"

- After each question, your child crosses off the numbers that have been eliminated.

- Your child wins the game if he or she guesses the secret number with up to 4 questions.

Number Puzzle

Work with your child to complete the number puzzle.

Across

1. 7 + 7 = ___

3. 170, 180, 190, ___

4. 1 + 2 + 3 + 4 + 5 = ___

6. 15, 17, 19, ___

Down

1. 8, 10, ___, 14

2. 100, 200, 300, ___

5. 40, 45, 50, ___

7. 8 + 8 = ___

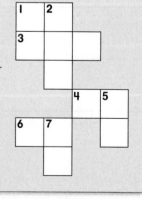

Repeating and Growing Patterns

NCTM Standards 2, 6, 7, 8, 9, 10
TEKS 2.6C, 2.14

What comes next? Continue each pattern.

1.

2.

X O X O X O X O X O X ___ ___ ___

3.

4 5 6 4 5 6 4 5 6 ___ ___ ___

4.

5. Make your own pattern. Draw it here.

 NOTE: Your child is learning about patterns that repeat and grow. Together, look around your home for different patterns.

     III   three   3

Is it a repeating pattern? Circle *yes* or *no*.
If *yes*, circle the pattern unit.

6. yes no

7. yes no

8.

5   10   15   20   25

yes no

9. yes no

 10. Choose a pattern from above that does NOT repeat.
How does the pattern grow?

_____

_____

_____

Problem Solving

11. Carey gets 5 cents each day.
In how many days will she have
25 cents? Use words, numbers,
or pictures to explain.

_____ days

4 four IV

Name _____ Date _____

Working with Number Patterns

NCTM Standards 1, 2, 6, 9, 10

TEKS 2.6A

1. Continue the pattern. What is missing?

Stair-Step Table	
Number of Steps Up	**Number of Dots**
1	1
2	3

How many dots are on each card?

2.

3. □

4. □

5. □

© Education Development Center, Inc.

🏠 **NOTE:** Your child is using Stair-Step Cards to show patterns. The cards can be put together to make squares and show designs.

  V five 5

Each figure is made from two cards. How many dots are in each figure? Record below.

6.

7.

8.

9.

10.

11.

12.

13.

14.

15.

16.

6	7	8	9	10	11	12	13	14	15	16

Challenge

How many dots are missing from each figure?

17.

18.

© Education Development Center, Inc.

Name _____   Date _____

Writing Number Sentences

NCTM Standards 1, 2, 6, 7, 8, 9, 10

TEKS 2.5C, 2.13A

Write number sentences to go with each figure.

I.

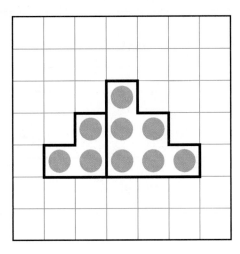

$$9 - 3 = 6$$

$$1 + 2 + 3 + 2 + 1 = 9$$

$$5 + 3 + 1 = 9$$

2.

_____

_____

_____

3.

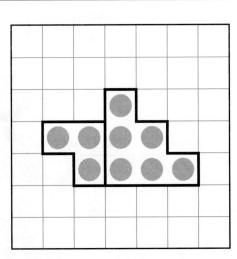

_____

_____

_____

© Education Development Center, Inc.

NOTE: Each figure is made from two Stair-Step Cards. Your child is learning to write number sentences about the dots in each figure by looking at the cards, the rows, and the columns.

 VII   seven   7

Write number sentences to go with each figure.

4.

_____

_____

_____

5.

_____

_____

_____

6.

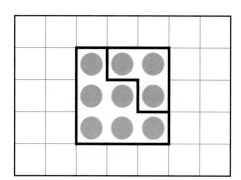

_____

_____

_____

Problem Solving

How are the sentences different?

7. How can both $3 + 6 = 9$ and $6 + 3 = 9$ tell about the same picture?

_____

_____

_____

Adding and Subtracting on the Number Line

NCTM Standards 1, 2, 6, 9, 10
TEKS 2.3A, 2.8

Draw the jump.

What is missing?

1.

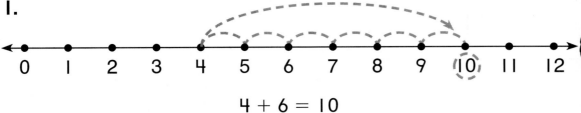

$$4 + 6 = 10$$

2.

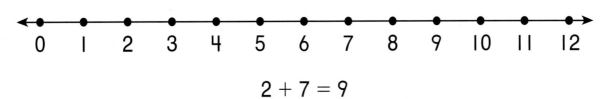

$$2 + 7 = 9$$

3.

$$8 + 3 = 11$$

4.

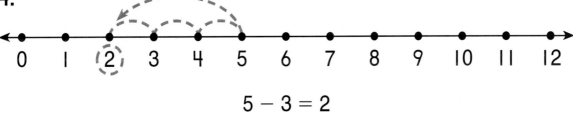

$$5 - 3 = 2$$

5.

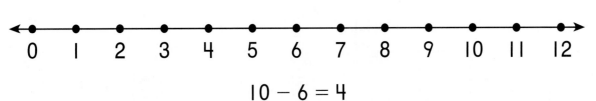

$$10 - 6 = 4$$

NOTE: Your child is using jumps on the number lines to show addition and subtraction.

   IX nine **9**

What number sentence is shown by the jump?

6.

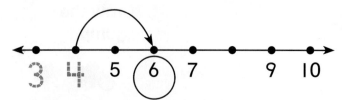

| 4 | + | 2 | = | 6 |

7.

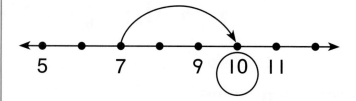

| ☐ | + | ☐ | = | ☐ |

8.

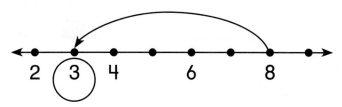

| ☐ | − | ☐ | = | ☐ |

9.

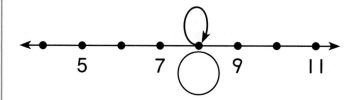

| ☐ | + | ☐ | = | ☐ |

10.

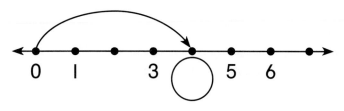

| ☐ | + | ☐ | = | ☐ |

11.

| ☐ | − | ☐ | = | ☐ |

Challenge
Make your own.

12.

0

| 0 | + | ☐ | = | ☐ |

13.

1

| ☐ | − | ☐ | = | ☐ |

Completing Number Sentences

NCTM Standards 1, 2, 6, 10
TEKS 2.3A, 2.8

What number is missing?

1.

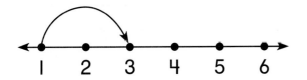

$$\boxed{1} + \boxed{2} = \boxed{3}$$

2.

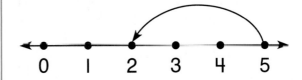

$$\boxed{5} - \boxed{3} = \boxed{}$$

3.

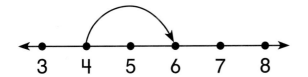

$$\boxed{} + \boxed{2} = \boxed{6}$$

4.

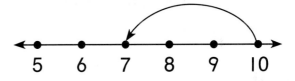

$$\boxed{} - \boxed{3} = \boxed{7}$$

5.

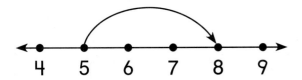

$$\boxed{5} + \boxed{} = \boxed{8}$$

6.

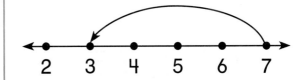

$$\boxed{7} - \boxed{} = \boxed{3}$$

NOTE: Your child is learning to relate jumps
on a number line to number sentences.
Each number sentence shows the starting
number, the jump size, and the landing number.

    XI eleven 11

What number is missing?

7.

6 7 8 9 10

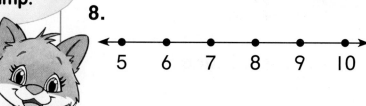

Now draw the jump.

8.

5 6 7 8 9 10

| 9 | + | | = | 10 |

| 10 | – | | = | 8 |

9.

4 5 6 7 8 9 10

10.

6 7 8 9 10 11 12

| | + | 3 | = | 10 |

| | – | 4 | = | 6 |

Challenge

11. Find as many ways as you can.

| | + | | = | 10 |

| | + | | = | 10 |

| | + | | = | 10 |

| | + | | = | 10 |

| | + | | = | 10 |

| | + | | = | 10 |

| | + | | = | 10 |

| | + | | = | 10 |

| | + | | = | 10 |

| | + | | = | 10 |

| | + | | = | 10 |

| | + | | = | 10 |

Name _____ Date _____

Skip-Counting on the Number Line

NCTM Standards 1, 2, 5, 6, 7, 8, 9, 10
🔺 TEKS 2.6A, 2.6B, 2.8, 2.14

Skip-count. What is missing?

1. Start at 0. The jump size is 2.

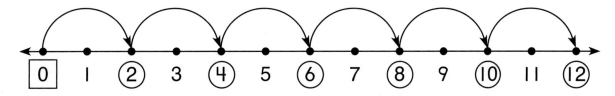

Number of Jumps	0	1	2	3	4	5	6	7
Landing Number	0	2	4	6				

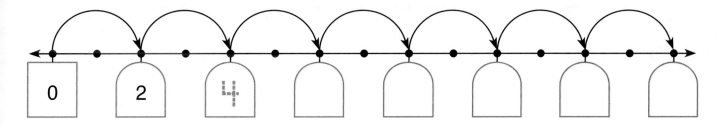

| 0 | 2 | 4 | | | | | |

---

2. Start at 1. The jump size is 2.

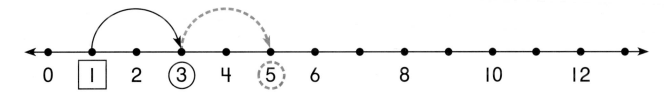

Number of Jumps	0	1	2	3	4	5	6	7
Landing Number	1	3	5					

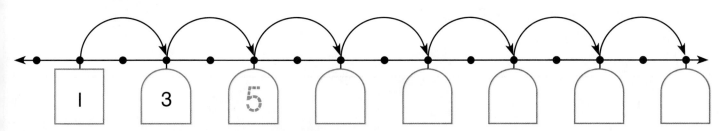

| 1 | 3 | 5 | | | | | |

NOTE: Your child is learning to skip-count by different numbers on the number line. Together, practice skip-counting by twos from different starting numbers.

XIII thirteen **13**

Skip-count. What are the missing numbers?

3.

4.

5.

6. Make your own.

Problem Solving

7. Gracie says she can start at 0 and skip-count to 12 by threes. Tal says he can start at 0 and skip-count to 12 by fours. Who is right? Explain.

_____

_____

Name _____ Date _____

More Skip-Counting on the Number Line

NCTM Standards 1, 2, 6, 7, 8, 9, 10
TEKS 2.3A, 2.3B, 2.8, 2.14

What is missing?

1.

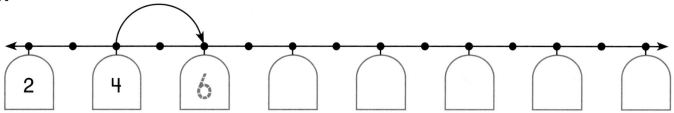

$$4 + \boxed{} = \boxed{6}$$

2.

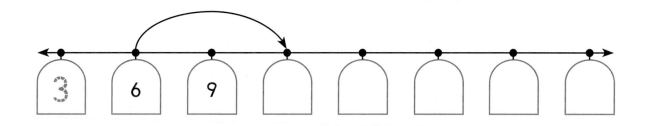

$$6 + \boxed{} = \boxed{}$$

3.

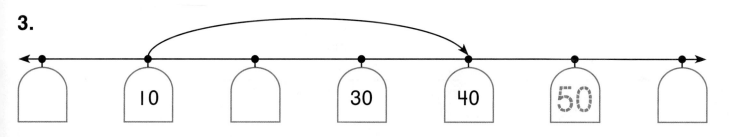

$$\boxed{} + \boxed{} = \boxed{}$$

 NOTE: Your child is learning to work with number-lines where the space between the dots is worth more than 1. The number lines can be used to help complete the addition sentences.

4. What is missing?

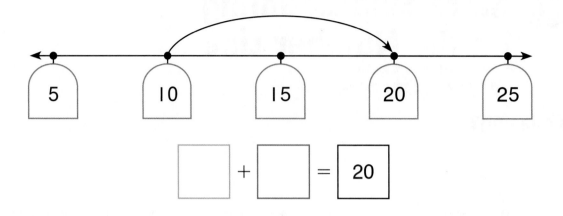

$$\boxed{} + \boxed{} = \boxed{20}$$

---

5. Make your own number line to show $20 + 8 = 28$.

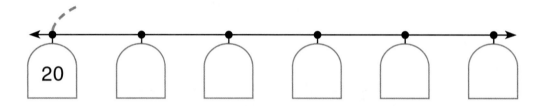

20

Problem Solving

6. You want to show $40 + 10 = 50$. What are some different ways to label the number line? Explain.

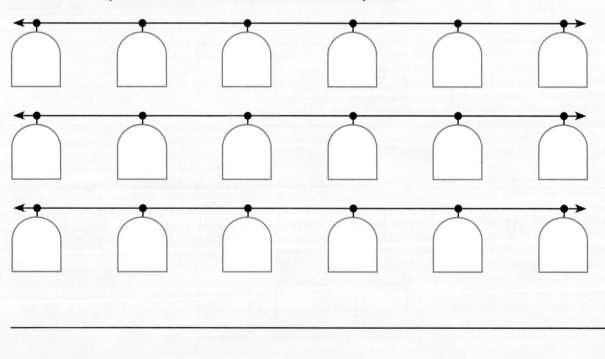

_____

_____

sixteen **XVI** $8 + 8$

Chapter 1
Lesson 8

Systematic Counting

NCTM Standards 1, 2, 6, 9, 10

TEKS 2.6A

How many different towers can you build?
Follow the rules. Color to show the towers.
Mark an X on towers you do not color.

Rules: • Use one blue cube in each tower.
         • Use cubes of another color to make the right height.

Height	Different Towers	Number of Towers
1. 1 cube tall		
2. 2 cubes tall		
3. 3 cubes tall		

NOTE: Your child is using cubes to build all possible towers
for each height. Only one of the cubes is blue and the rest
are another color.

How many different towers can you build?

Height	Different Towers	Number of Towers
4. 4 cubes tall		_____
5. 5 cubes tall		_____
6. 6 cubes tall		_____

Challenge

7. What is missing?

Number of Cubes	1	2	3					
Number of Different Towers								

Finding Ways to Make 10

NCTM Standards 1, 2, 6, 7, 8, 9, 10

🔻 TEKS 2.3A, 2.14

Sums of 10 Search

1. Which pairs make 10? Circle them as fast as you can.

4	6	6	8	9	1	1	5	2
6	4	5	2	1	8	9	6	8

3	3	8	9	6	2	2	4	5
6	7	2	1	3	9	8	6	5

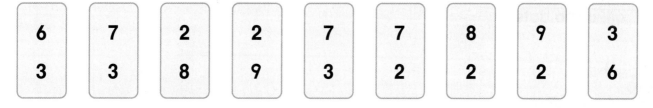

6	7	2	2	7	7	8	9	3
3	3	8	9	3	2	2	2	6

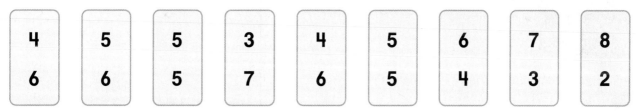

4	5	5	3	4	5	6	7	8
6	6	5	7	6	5	4	3	2

9	2	7	7	8	3	4	5	1
1	9	2	3	2	8	7	5	8

0	8	6	7	4	9	6	1	3
9	2	4	3	6	1	3	9	7

NOTE: Your child is learning to quickly recognize pairs of numbers with a sum of 10. You can practice by saying a number and having your child name that number's partner to make 10.

2. How many ways can you make 10?

Cover some, all, or no dots.

Uncovered	Covered
3	7

Problem Solving

3. I am thinking of two numbers with a sum of 10. One of the numbers is even. What can you say about the other number? What two numbers might they be?

_____

_____

Previewing Multiplication, Part I

NCTM Standards 1, 2, 3, 6, 9, 10
TEKS 2.13A, 2.14

How many intersections are there?
Write the missing numbers.

An intersection
is where two
lines meet.

1.

2.

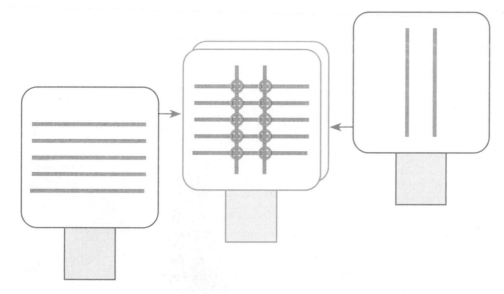

NOTE: Your child is exploring and counting
intersecting lines. Pictures like these will be
used later to model multiplication.

20 + 1   ▲   **XXI**   twenty-one   21

What is missing?

3.

4.

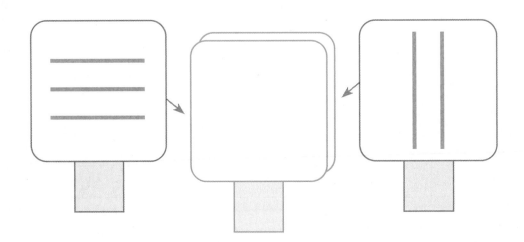

Challenge

5. Make your own.

What will you do first?

Chapter 1
Lesson 11

Previewing Multiplication, Part II

NCTM Standards 1, 2, 3, 6, 9, 10
TEKS 2.13A

What words can you make?

1.

2.

3.

4.

5.

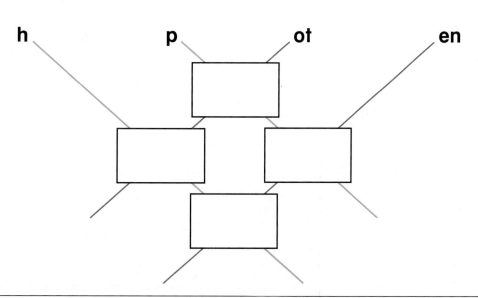

NOTE: Your child is learning to make words by combining letters at the intersections of lines. Pictures like these will be used later to model multiplication.

6. What words can you make?

a   i   s   n   t

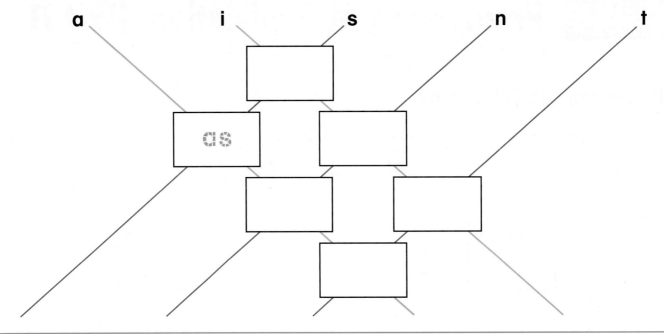

as

7. Make your own.

beginning   ending   ending

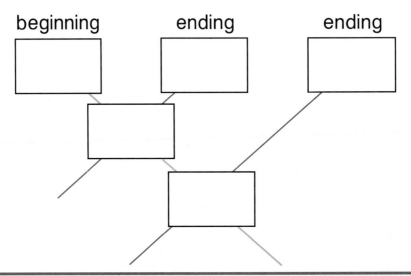

Problem Solving

8. Kermit is making sandwiches with one meat and one cheese. Write a list of all the different sandwiches he can make.

Meat	Cheese
bologna	American
turkey	Swiss

_____ _____

_____ _____

_____ _____

Problem Solving Strategy
Look for a Pattern

NCTM Standards 1, 2, 6, 7, 8, 9, 10
🪧 **TEKS 2.6C, 2.12B, 2.12C**

Understand
Plan
Solve
Check

I. What is the number of the third house after 18 South Street? Explain.

South Street

12   14   16   18

_____

_____

_____

2. What is the number of the house where the next tree should be planted? Explain.

South Street

55   57   59   61   63

_____

_____

_____

3. Tammy is building a fence around her yard. Draw the next two fence posts. Explain how you know what to draw.

_____

_____

 NOTE: Your child is exploring different ways to solve problems. Sometimes looking for a pattern is an efficient way to solve a problem.

Problem Solving Test Prep

1. Carla has 6 games. Jeff has 4 games. How many games do they have altogether?

   (A) 2 games

   (B) 6 games

   (C) 10 games

   (D) 24 games

2. There are 11 children on the playground. 8 are on the swings. The rest are playing catch. Which number sentence shows how many children are playing catch?

   (A) $11 + 8 = 19$

   (B) $11 - 8 = 3$

   (C) $8 + 8 = 16$

   (D) $11 - 4 = 7$

Show What You Know

3. Matt bakes 15 muffins. He gives some to his mother. He gives 3 fewer muffins to his brother. He has 6 muffins left. How many muffins did he give to his mother?

   _____ muffins

   Explain how you found the answer.

   _____

   _____

   _____

   _____

   _____

   _____

4. The highest temperature on Monday was 47°. The temperature went up 2° each day. On what day will the highest temperature be 53°?

   _____

   Explain how you found the answer.

   _____

   _____

   _____

   _____

   _____

   _____

Chapter 1 # Review/Assessment

NCTM Standards 1, 2, 3, 6, 7, 8, 9, 10

1. Continue the pattern. Lesson 1

 _____

2. How many dots are in each figure? Lesson 2

  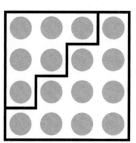

_____   _____   _____

Write number sentences to go with each figure. Lesson 3

3.

_____

_____

_____

4.

_____

_____

_____

What number sentence is shown by the jump? Lesson 4

5.

_____

6.

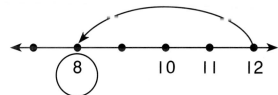

_____

What number is missing? Lesson 5

7.

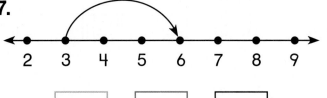

| 3 | + | 3 | = | |

8.

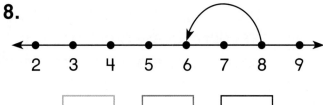

| | − | 2 | = | 6 |

What is missing? <small>Lessons 6, 7</small>

9.

10.

$$5 + \boxed{} = \boxed{}$$

11. Circle pairs with a sum of 10. <small>Lesson 9</small>

8	6	1	3	5	4	3	2	9	5
2	5	9	7	5	7	6	8	1	5

12. How many intersections are there? <small>Lesson 10</small>

3   4

Problem Solving <small>Lesson 12</small>

13. At Mt. Way School, there are 7 doors on one side of the hallway. Starting with the first door, every other one is painted blue. How many blue doors are there?

_____ blue doors

Chapter 2

Working with 10
Stories About 10 ✏

🔺 **TEKS 2.3A**

Take 10 counters. Separate them into two sets.

STEP 1 Counting Sets

How many counters do you have in all? _____

Does the total change when you move them around? _____

Explain your answer. _____

_____

_____

STEP 2 Telling Stories

Write a number story about your two sets.
Draw a picture to show the story.

_____

_____

STEP 3 Recording Sentences

What number sentence can describe the sets?

_____

What other number sentence can describe the same sets?

_____

Investigation

Dear Family,

Today we started Chapter 2 of *Think Math!* In this chapter, I will work with number pairs that have a sum of 10 (such as 4 + 6). I will also start at any number and add or subtract 10 and develop strategies to look at a number and tell how much more or less than 10 it is. There are NOTES on the Lesson Activity Book pages to explain what I am learning every day.

Here are some activities for us to do together at home. These activities will help me practice working with 10.

Love,

_____

Family Fun

Fingers Up for 10!

Work with your child to add two numbers by making 10.

- Name two numbers from 5 to 10. Show one number with your fingers. Your child shows the other number. Make sure to use all of the fingers of one hand for the first 5 of each number.

- Add the two numbers by making 10. Put your hand of 5 with your child's hand of 5. This makes 10. Then add the rest of the fingers and add it to 10.

- Repeat several times with different pairs of numbers from 5 to 10.

X-Concentration

Work with your child to practice adding two numbers with a sum of 10.

- Gather a deck of playing cards and remove the 10s, jacks, queens, and kings.

- Mix the 36 cards. Place them face down in 4 rows of 9 like this.

- Take turns with your child. On each turn, a player turns two cards face up. If their sum is 10, the player takes the cards and goes again. The player continues until the pair does not add up to 10.

- Take turns until all cards are paired.

- As a variation, play with cards from only 2 suits.

Chapter 2
Lesson 1

Finding Sums of 10

NCTM Standards 1, 2, 6, 7, 8, 9, 10
🔺 **TEKS 2.3A**

1. Find different ways to put 10 counters
in two sets. Record below.

| 0 | 10 | 0 + 10 = 10 |

_____ + _____ = 10

_____ + _____ = 10

_____ + _____ = 10

**Use the same
10 counters
each time.**

_____ + _____ = 10

_____ + _____ = 10

_____ + _____ = 10

_____ + _____ = 10

_____ + _____ = 10

_____ + _____ = 10

_____ + _____ = 10

_____ + _____ = 10

🏠 **NOTE:** Your child is learning all of the
different addition sentences with a
sum of 10.

What is missing?

2.

_____ + _____ = _____

3.

_____ + _____ = _____

4.

_____ + _____ = _____

5.

_____ + _____ = _____

Challenge

What is the value of in each problem?

6.

_____ + 2 = 10

7.

_____ + 4 = 10

Introducing > and <

NCTM Standards 1, 2, 6, 7, 8, 9, 10
TEKS 2.1C, 2.14

Write >, <, or =. Make each sentence true.

1. 1 + 4 ◯> 1 + 3

2. 5 + 1 ◯ 5 + 2

3. 2 + 8 ◯ 1 + 7

4. 5 + 6 ◯ 6 + 5

5. 1 + 8 ◯ 1 + 6

6. 3 + 6 ◯ 2 + 6

7. 2 + 3 ◯ 3 + 2

8. 7 + 3 ◯ 3 + 7

9. 4 + 2 ◯ 2 + 8

10. 6 + 5 ◯ 5 + 5

11. 4 + 5 ◯ 1 + 3

12. 6 + 5 ◯ 7 + 5

NOTE: Your child is learning to compare addition expressions using the symbols <, >, or = to make each sentence true. < means "is less than", > means "is greater than", and = means "is equal to."

11 + 11 + 11   **33**   **XXXIII**   thirty-three   33

What is missing? Make each sentence true.

13. $4 + 5 < 4 +$ _____

14. $8 + 2 < 8 +$ _____

15. $3 + 2 >$ _____ $+ 2$

16. $5 + 3 >$ _____ $+ 3$

17. $4 + 5 <$ _____ $+ 5$

18. $1 + 6 > 1 +$ _____

19. $2 + 7 >$ _____ $+$ _____

20. $3 + 9 >$ _____ $+$ _____

21. $7 + 8 <$ _____ $+$ _____

22. $10 + 9 <$ _____ $+$ _____

23. What addition expressions make the sentence true?

$7 + 4 >$ _____ $+$ _____

_____ $+$ _____   _____ $+$ _____   _____ $+$ _____

_____ $+$ _____   _____ $+$ _____   _____ $+$ _____

Problem Solving

24. Ned has two stools. The metal stool is 5 inches tall. The wooden stool is 14 inches tall. On which stool can Ned stand higher? Explain.

_____

_____

Finding and Comparing Sums to 10

NCTM Standards 1, 2, 8, 9, 10
TEKS 2.1C, 2.3A

Sums Greater than 10 Search

1. Which pairs make sums greater than 10?
Circle them as fast as you can.

5 5	5 6	5 7	4 4	4 6	7 4	7 3	4 7	2 8

3 7	3 5	3 8	8 1	6 4	6 7	7 2	6 5	4 6

4 7	2 7	7 3	9 0	1 9	2 9	1 8	3 6	6 5

0 10	7 5	9 1	4 5	0 9	8 3	8 2	9 2	10 0

NOTE: Your child is learning to identify number pairs with sums greater than 10 and compare addition problems to 10 or 20.

Write >, <, or =.

2. $10 \bigcirc< 7 + 5$

3. $20 \bigcirc 17 + 5$

4. $2 + 9 \bigcirc 10$

5. $12 + 9 \bigcirc 20$

6. $8 + 4 \bigcirc 10$

7. $8 + 14 \bigcirc 20$

8. $10 \bigcirc 3 + 7$

9. $20 \bigcirc 3 + 17$

10. $5 + 6 \bigcirc 2 + 8$

11. $15 + 6 \bigcirc 20$

12. $2 + 7 \bigcirc 10$

13. $2 + 17 \bigcirc 20$

Challenge
Write >, <, or =.

14. If $6 + 7 \bigcirc 10$, then $16 + 17 \bigcirc 30$.

15. If $8 + 2 \bigcirc 10$, then $18 + 12 \bigcirc 30$.

Chapter 2
Lesson 4

Mastering Sums of 10

NCTM Standards 1, 2, 6, 7, 8, 9, 10
TEKS 2.1C, 2.3A, 2.14

Write >, <, or =. Make each sentence true.

1. $4 + 6 \bigcirc > \bigcirc 4 + 5$

2. $9 + 2 \bigcirc 1 + 9$

3. $1 + 8 \bigcirc 1 + 9$

4. $3 + 6 \bigcirc 4 + 6$

5. $4 + 6 \bigcirc 3 + 7$

6. $1 + 8 \bigcirc 2 + 8$

7. $3 + 7 \bigcirc 7 + 4$

8. $4 + 6 \bigcirc 2 + 8$

9. $7 + 3 \bigcirc 5 + 4$

10. $1 + 3 \bigcirc 1 + 9$

11. $8 + 3 \bigcirc 2 + 8$

12. $5 + 5 \bigcirc 6 + 5$

NOTE: Your child is memorizing the facts with a sum of 10. These facts will help him or her later to solve other facts.

What is missing? Make each sentence true.

13. $4 + 6 = 3 +$ _____

14. $5 + 5 = 8 +$ _____

15. $1 + 9 =$ _____ $+ 8$

16. $7 + 3 =$ _____ $+ 7$

17. $5 + 5 <$ _____ $+ 5$

18. $9 + 1 > 1 +$ _____

19. $6 + 4 <$ _____ $+ 9$

20. $0 + 10 > 4 +$ _____

21. _____ $+ 4 < 3 + 7$

22. _____ $+ 6 > 3 + 7$

23. Make the sentence true. Explain your answer.

$2 + 8 < 2 +$ _____

_____

_____

_____

_____

Challenge

24. How can you make four number pairs with sums of 10? Draw lines to match the numbers below.

1   2   3   4   5   6   7   8   9

Name _____ Date _____

Adding and Subtracting with 10

NCTM Standards 1, 2, 6, 7, 8, 9, 10
TEKS 2.3B, 2.5A

Add 10. What number is missing?

1. $51 + 10 =$ | 61 |

2. $4 + 10 =$ [ ]

3. $32 + 10 =$ [ ]

4. [ ] $+ 10 = 79$

5. $28 + 10 =$ [ ]

6. [ ] $+ 10 = 26$

Move up exactly one row.

90	91	92	93	94	95	96	97	98	99
80	81	82	83	84	85	86	87	88	89
70	71	72	73	74	75	76	77	78	79
60	61	62	63	64	65	66	67	68	69
50	51	52	53	54	55	56	57	58	59
40	41	42	43	44	45	46	47	48	49
30	31	32	33	34	35	36	37	38	39
20	21	22	23	24	25	26	27	28	29
10	11	12	13	14	15	16	17	18	19
0	1	2	3	4	5	6	7	8	9

7.

20			77			95		
10	20	24	67	58	82	85	90	46

8. $85 + 10 =$ [ ]

9. [ ] $+ 10 = 20$

10. $90 + 10 =$ [ ]

11. [ ] $+ 10 = 56$

12. $20 + 10 =$ [ ]

13. $24 + 10 =$ [ ]

14. $82 + 10 =$ [ ]

15. $58 + 10 =$ [ ]

16. [ ] $+ 10 = 77$

NOTE: Your child is learning to use a hundred grid to add and subtract 10 from any two-digit number.

Subtract 10. What number is missing?

17. 72 − 10 = 62

18. 30 − 10 = ☐

19. 91 − 10 = ☐

20. ☐ − 10 = 33

21. ☐ − 10 = 14

22. 85 − 10 = ☐

Move down exactly one row.

90	91	92	93	94	95	96	97	98	99
80	81	82	83	84	85	86	87	88	89
70	71	72	73	74	75	76	77	78	79
60	61	62	63	64	65	66	67	68	69
50	51	52	53	54	55	56	57	58	59
40	41	42	43	44	45	46	47	48	49
30	31	32	33	34	35	36	37	38	39
20	21	22	23	24	25	26	27	28	29
10	11	12	13	14	15	16	17	18	19
0	1	2	3	4	5	6	7	8	9

23.

24	29	72	85	17	43	30	91	90
14		62				20		

Add and subtract.

24. 64 + 10 = ☐

64 − 10 = ☐

25. 56 + 10 = ☐

☐ − 10 = 46

26. ☐ + 10 = 48

☐ − 10 = 28

Challenge
What is the missing number?

27.

	57	

28.

		34		

40 forty **XL 8**

Finding How Close to 10

NCTM Standards 1, 2, 6, 7, 8, 9, 10
TEKS 2.3A, 2.12A, 2.12B, 2.14

How close to 10 is the sum?

1. 7 + 1

10 | -- 2

2. 4 + 7

10 | ⬜

3. 6 + 6

10 | ⬜

4. 5 + 4

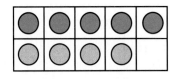

10 | ⬜

5. 8 + 2

10 | ⬜

6. 3 + 5

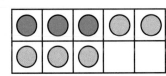

10 | ⬜

7. 3 + 9

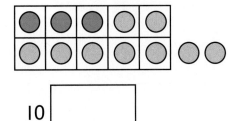

10 | ⬜

8. 1 + 9

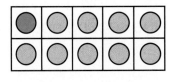

10 | ⬜

NOTE: Your child is learning to identify how close a sum is to 10. Ask him or her to explain the exercises on this page.

Make each sentence true.

9. $8 + 3 = 10$

10. $2 + 7 = 10$

11. $4 + 4 = 10$

12. $4 + 8 = 10$

13. $9 + 3 = 10$ ☐

14. $8 + 1 = 10$ ☐

Make your own.

15. ☐ + ☐ = $10 + 1$

16. ☐ + ☐ = $10 - 2$

Problem Solving

17. Jill buys two packs of 10 hot dogs each. She serves 6 hot dogs on Friday and 6 on Saturday. How many hot dogs does she use from the second pack? Explain.

_____

_____

Chapter 2
Lesson 7

Adding Numbers by Making 10

NCTM Standards 1, 2, 6, 7, 8, 9, 10
🔻 TEKS 2.3A, 2.14

Rewrite each fact with a 10.

Make 10 with 5 from each number. Then add what is left over.

1. $6 + 8 = \boxed{14}$

$10 + 4 = 14$

2. $7 + 9 = \boxed{}$

10

3. $8 + 5 = \boxed{}$

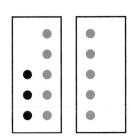

_____

4. $9 + 8 = \boxed{}$

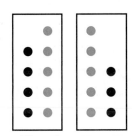

_____

5. $8 + 7 = \boxed{}$

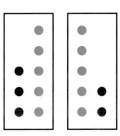

_____

6. $9 + 9 = \boxed{}$

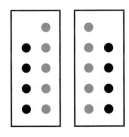

_____

7. $7 + 6 = \boxed{}$

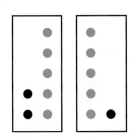

_____

8. $6 + 9 = \boxed{}$

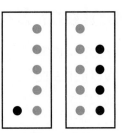

_____

NOTE: Your child is learning a strategy to add two numbers by making 10 with a 5 from each number and then adding the remaining numbers to 10.

Add fives to find each sum. What is missing?

9. 6 + 9 =

5 + **1** = 6

5 + **4** = 9

10 + **5** = **15**

10. 8 + 7 =

5 + ___ = 8

5 + ___ = 7

10 + ___ = ☐

11. 9 + 5 =

5 + ___ = 9

5 + ___ = 5

10 + ___ = ☐

12. 7 + 6 =

___ + ___ = 7

___ + ___ = 6

___ + ___ = ☐

13. 8 + 9 =

___ + ___ = 8

___ + ___ = 9

___ + ___ = ☐

14. 6 + 8 =

___ + ___ = 6

___ + ___ = 8

___ + ___ = ☐

15. How many different ways can you solve 7 + 9?
Use words, numbers, or pictures to explain.

Problem Solving

16. Kat made 10 to solve 9 + 6. She did not add fives.
What strategy might she have used?

_____

_____

Chapter 2
Lesson 8

Rounding to the Nearest 10

NCTM Standards 1, 2, 6, 7, 8, 9, 10
TEKS 2.1C, 2.8, 2.14

Write the tens that sandwich the number on each number line. Which is the nearest ten?

1.

| 80 | | | | | | | | 88 | | 90 |

The nearest ten to 88 is ___90___.

2.

| | | | 24 | | | | | | | |

The nearest ten to 24 is _____.

3.

| | | | | 35 | | | | | | |

The nearest ten to 35 is _____.

4.

| | | | | | | 67 | | | |

The nearest ten to 67 is _____.

 5. What is the rule for rounding to the nearest ten? _____

_____

_____

NOTE: Your child is learning a strategy to round two-digit numbers to the nearest multiple of ten. This skill may be used in various ways throughout mathematics.

 9   **XLV** forty-five 45

Make a sandwich with each number line. What number might come between the tens?

Write the number in position on the number line.

6. The nearest ten is 30.

20   [28]   30

7. The nearest ten is 50.

50   60

8. The nearest ten is 80.

80   90

9. The nearest ten is 40. Write all possible numbers that could be rounded to 40.

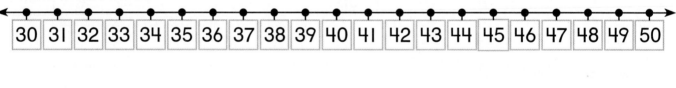

30 31 32 33 34 35 36 37 38 39 40 41 42 43 44 45 46 47 48 49 50

_____   _____   _____   _____   _____

_____   _____   _____   _____

Problem Solving

70¢

10. Daunte wants to buy an apple for 70¢. He has 67¢.
He says he has enough money because 67 rounds to 70.
Is he right? Use words, numbers, or pictures to explain.

_____

_____

_____

Name _____ Date _____

Problem Solving Strategy
Solve a Simpler Problem ✏

NCTM Standards 1, 2, 4, 6, 7, 8, 9, 10
TEKS 2.12B, 2.12C, 2.14

Understand
Plan
Solve
Check

1. Karen made 2 dozen cookies. Jim made 4 fewer cookies than Karen. How many cookies did they make altogether?

_____ cookies

How did you find the answer?

2. Karen gave 8 cookies to Sasha and 5 to her brother. Jim gave 9 cookies to Oleg and 6 to his sister. Who gave away more cookies, Karen or Jim?

_____

How did you find the answer?

3. There are 53 tiles on the floor. It takes 1 minute to paint each tile. How many minutes will it take to paint the entire floor?

_____ minutes

How did you find the answer?

NOTE: Your child is exploring different ways to solve problems. Sometimes solving a simpler problem is an efficient way to solve a problem.

Problem Solving Test Prep

1. Sue practices her violin for 3 hours each day. In how many days will she practice for a total of 12 hours?

   Ⓐ 36 days

   Ⓑ 15 days

   Ⓒ 9 days

   Ⓓ 4 days

2. Cliff has 2 coins. Both coins are the same. Which amount could he NOT have as the total?

   Ⓐ 10¢

   Ⓑ 20¢

   Ⓒ 40¢

   Ⓓ 50¢

Show What You Know

3. Raj, Bev, and Al are running a race. One person comes in first place, one in second, and one in third place. How many different ways can they finish the race?

   _____ ways

   Explain how you found the answer.

   _____

   _____

   _____

   _____

   _____

   _____

4. Pia saw birds and dogs at the pet store. She saw 4 heads and 12 feet. How many of each animal did she see?

   _____ birds _____ dogs

   Explain how you know.

   _____

   _____

   _____

   _____

   _____

   _____

Chapter 2 Review/Assessment

NCTM Standards 1, 2, 6, 7, 8, 9, 10

What is missing? Lesson 1

1.

_____ + _____ = _____

2.

_____ + _____ = _____

Write >, <, or =. Lessons 2, 3

3. $1 + 5$ ◯ $1 + 2$

4. $10 + 6$ ◯ 20

What is missing? Make each sentence true. Lesson 4

5. $8 + 2 < 3 +$ _____

6. _____ $+ 3 < 4 + 6$

Add or subtract 10. What number is missing? Lesson 5

7. $68 + 10 =$ _____

8. _____ $+ 10 = 33$

9. $75 - 10 =$ _____

10. _____ $- 10 = 41$

90	91	92	93	94	95	96	97	98	99
80	81	82	83	84	85	86	87	88	89
70	71	72	73	74	75	76	77	78	79
60	61	62	63	64	65	66	67	68	69
50	51	52	53	54	55	56	57	58	59
40	41	42	43	44	45	46	47	48	49
30	31	32	33	34	35	36	37	38	39
20	21	22	23	24	25	26	27	28	29
10	11	12	13	14	15	16	17	18	19
0	1	2	3	4	5	6	7	8	9

How close to 10 is the sum? Lesson 6

11. $3 + 8$

10 ☐

12. $6 + 4$

10 ☐

Rewrite each fact with a 10. Lesson 7

13. $9 + 6 = ?$

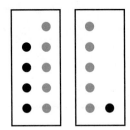

_____

14. $7 + 8 = ?$

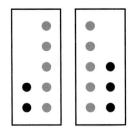

_____

15. $8 + 5 = ?$

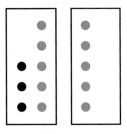

_____

16. What number might come between the tens?
The nearest ten is 70. Lesson 8

60   70

Problem Solving Lesson 9

17. Maya's grade at school has 90 students.
Her brother's grade has 70 students.
How many students are in the two grades together? _____ students

How did you find the answer?

Chapter 3

Place Value
Counting Counters ✏️

Look at the counters in your group's bag.

STEP 1 Estimating

How many counters do you think are in your bag? _____

How did you decide? _____

_____

STEP 2 Counting

What is the exact number of counters in your bag? _____

How did you find the exact number? _____

_____

_____

STEP 3 Counting Another Way

What is another way to find the exact number? _____

_____

Which way do you like better? Explain.

_____

_____

_____

Investigation

Dear Family,

Today we started Chapter 3 in *Think Math!* In this chapter, I will estimate, count, read and write numbers, and draw symbols for numbers to 999. There are NOTES on the Lesson Activity Book pages to explain what I am learning every day.

Here are some activities for us to do together at home. These activities will help me understand place value.

Love,

_____

Family Fun

Place-Value Match

Work with your child to prepare game cards to play *Place-Value Match*.

- Use index cards or slips of paper to make a set of 32 game cards. On 16 cards, write a three-digit number. On the remaining cards, write the number of hundreds, tens, and ones in each three-digit number.

- Partners shuffle the cards, and each one takes 6 cards. They put the remaining cards in a facedown stack.

- Players take turns asking each other for cards to make a match. For example:

    Do you have a card for 4 hundreds, 7 tens, 3 ones?

    If the partner does not have the matching card, the player chooses a card from the stack. If the partner has the card, a match is made, and the player puts the pair of cards aside.

- The first player to match all of his or her cards wins the game.

Guess How Many?

Work with your child to estimate and count objects.

- Gather a container of small identical objects such as pennies, paper clips, or marbles. Ask your child to take a handful of the objects and count them.

- Without counting, each player guesses the number of objects in the container and writes the guess on a sheet of paper.

- Players then work together to group the objects into sets of ten to count the total number of objects. The objects are returned to the container.

- The player with the guess closer to the actual number is the winner.

Chapter 3
Lesson 1

Estimating and Counting Larger Numbers

NCTM Standards 1, 6, 7, 8, 9, 10
⬆ **TEKS 2.13A, 2.14**

How many are there? Use the set of 10 to estimate. Then count to find the total.

1.

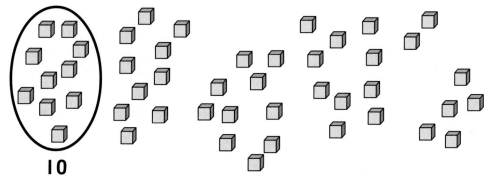

10

45	47
Estimate	Count

2.

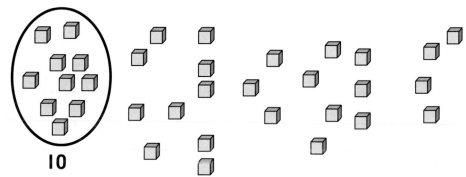

10

_____	_____
Estimate	Count

3.

10

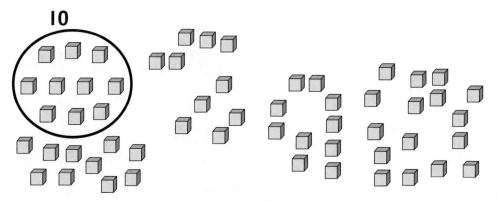

_____	_____
Estimate	Count

 NOTE: Your child is learning to estimate by using 10 as a benchmark and is making smaller sets to count a large collection of objects.

4. Draw as many squares as you can in the frame.
Ask a classmate to estimate the total and make
sets to count.

Estimate _____ Count _____

Challenge

5. How many are there? Estimate.
Count some and revise your estimate.
Count it all to find the total.

Estimate _____

Revised Estimate _____

Total _____

Name _____ Date _____

Grouping by Tens and Hundreds

NCTM Standards 1, 6, 7, 8, 9, 10

TEKS 2.1A, 2.1B, 2.12A, 2.12B, 2.12C, 2.12D, 2.13A, 2.14

It is easier to count the wheels in sets of 10.

**Both pictures show the same number.
How many wheels do they show?**

1.

   ☐ ones

   5   7   ⇨   57
   tens ones

2.

   ☐ ones

   ____ ____ ⇨ _____
   tens ones

3.

   ☐ ones

   ____ ____ ⇨ _____
   tens ones

4.

   ☐ ones

   ____ ____ ⇨ _____
   tens ones

 NOTE: Your child is learning that it is sometimes easier to group objects by tens and hundreds when counting a large collection of objects.

5. All three pictures show the same number.
How many wheels do they show?

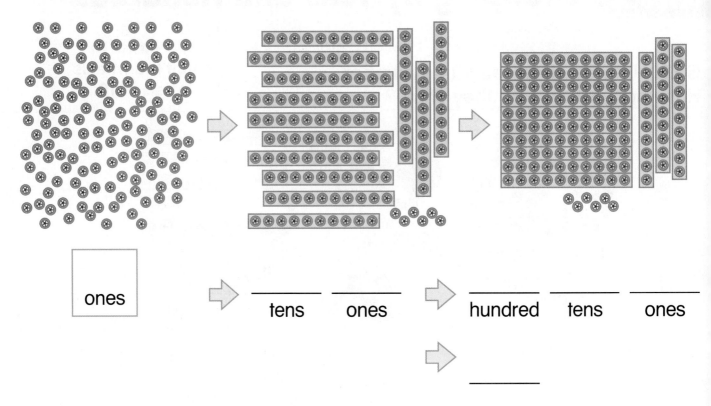

ones

⇨ ___ ___
   tens ones

⇨ ___ ___ ___
   hundred tens ones

⇨ ___

---

6. Circle the set in Problem 5 that is easiest to count.
Explain why you think this is so.

_____

_____

_____

Problem Solving

7. A millipede is a small animal with many
legs. This millipede has 115 pairs of legs.
How many legs does the millipede have?
(Hint: A pair is two.)

_____ hundreds _____ tens _____ ones = _____ legs

Name _____ Date _____

Representing Two-Digit Numbers

NCTM Standards 1, 6, 8, 9, 10
TEKS 2.1B, 2.12B, 2.12C, 2.13A, 2.14

What is missing? Draw symbols and write numbers.

1.

_____4_____ tens _____5_____ ones = ___40___ + ___5___ = ___45___

2.

_____ tens _____ ones = _____ + _____ = _____

3.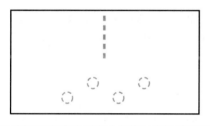

_____ tens 4 ones = 10 + _____ = _____

4.

_____ tens _____ ones = _____ + _____ = _____

5.

5 tens 3 ones = _____ + _____ = _____

6. Make your own.

_____ tens _____ ones = _____ + _____ = _____

NOTE: Your child is learning the value of the digits in
two-digit numbers. Ask your child to tell how many
tens and ones are in different numbers around you.

What number from the box will complete each riddle?

7. My tens digit has a value of 40.
   My ones digit is more than 5.

   49
   _____

15	19	22
28	32	35
43	49	53
69	72	85
88	90	94

8. My tens digit has a value of 90.
   My ones digit is less than 4.

   _____

9. Both of my digits are the same number. My tens digit has a value of 80.

   _____

10. My tens digit has a value less than 50. My ones digit has a value of 5.

    _____

11. My tens digit has a value less than 40. My ones digit is more than 7.

    _____

✏ 12. Make up your own riddle about a number in the box.

_____

_____

_____

Problem Solving

13. I have more than 4 tens.
    I have between 6 and 9 ones.

    What number could I be? _____
    Draw a picture of the number.

Chapter 3
Lesson 4

Representing Three-Digit Numbers

NCTM Standards 1, 6, 8, 9, 10
🔻 **TEKS 2.1B, 2.13A, 2.14**

What is missing? Draw symbols and write numbers.

1.

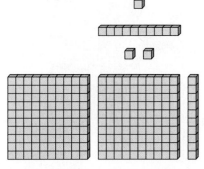

_2_ hundreds _2_ tens _3_ ones

_200_ + _20_ + _3_ = _223_

2.

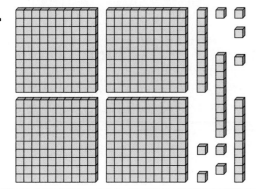

_____ hundreds _____ tens _____ ones

_____ + _____ + _____ = _____

3.

_____ hundreds _____ tens _____ ones

_____ + _____ + _____ = 706

4. Make your own.

_____ hundreds _____ tens _____ ones

_____ + _____ + _____ = _____

🏠 **NOTE:** Your child is learning the value of digits in three-digit numbers. Ask your child to tell how many hundreds, tens, and ones are in different numbers you name.

Toss 5 small counters onto the grid.
Record the numbers. An example is
given in Problem 5.

1	100	10	10	1	10	1
100	1	1	10	10	100	1
1	100	10	10	10	100	100

	Numbers That Counters Land On	Total
5.	10 + 100 + 1 + 1 + 10	122
6.	_____ + _____ + _____ + _____ + _____	_____
7.	_____ + _____ + _____ + _____ + _____	_____
8.	_____ + _____ + _____ + _____ + _____	_____

Challenge
What number is 10 more?

9. _____

10. _____

Chapter 3
Lesson 5

Regrouping

NCTM Standards 1, 6, 7, 8, 9, 10

 TEKS 2.1B, 2.12A, 2.12B, 2.13A, 2.14

Show each number with the fewest blocks.
Draw symbols for the blocks.

1.

134

2.

_____

3.

_____

4.

_____

5.

_____

6.

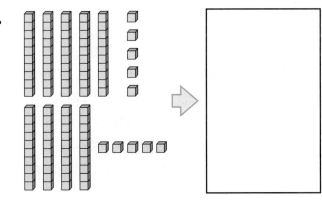

_____

NOTE: Your child is learning to regroup tens and ones to show the same number in different ways.

Regroup in different ways.

7.

24

_____0_____ tens _____24_____ ones

_____1_____ tens _____14_____ ones

_____ tens _____ ones

8.

38

_____ tens _____ ones

_____ tens _____ ones

_____ tens _____ ones

_____ tens _____ ones

 9. Choose any two-digit number. How many ways can you regroup? Draw or write to show the ways.

Number _____

Problem Solving

10. Elva has 36¢. She has only dimes and pennies. What different combinations of coins might she have? Use words, numbers, or pictures to explain.

Using Place Value to Compare

NCTM Standards 1, 2, 6, 7, 8, 9, 10
TEKS 2.1B, 2.1C, 2.12A, 2.12B, 2.13A, 2.14

Write the numbers. Then write >, <, or =.

1.

67 ⬜

2.

⬜ ⬜

3.

⬜ ⬜

4. Make your own.

⬜ ◯ ⬜

Write a digit to make each sentence true.

5. 678 < _7_ 78	6. 364 < ____ 64	7. 420 > ____ 20
8. 7 ____ 8 < 768	9. 8 ____ 4 > 864	10. 2 ____ 7 < 267
11. 40 ____ > 403	12. 26 ____ < 269	13. 51 ____ < 515
14. 623 > 6 ____ 3	15. ____ 92 > 492	16. 134 < 13 ____

 17. Write all possible answers for Problem 16.
Use words, numbers, or pictures to explain.

____   ____   ____

____   ____

Problem Solving

18. Jim's soccer team got 32 goals this season. The number of goals last season was a lot less. It had the same ones digit. How many goals do you think they got last year? Explain.

Use what you know about place value to help you.

_____

_____

_____

_____

Chapter 3
Lesson 7

Connecting Numbers and Words

NCTM Standards 1, 2, 6, 7, 8, 9, 10
TEKS 2.1C, 2.13A, 2.14

What number matches each word?

1. twenty-nine .... __29__

2. twelve .... _____

3. sixty....................... _____

4. one hundred ten............. _____

5. thirty-three .... _____

6. ninety-one .... _____

7. two hundred three .... _____

8. eighty-four .... _____

---

9. What is the order from smallest to biggest?

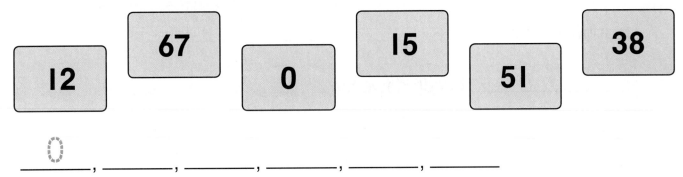

67   15   38
12   0   51

__0__, _____, _____, _____, _____, _____

---

10. What is the order from biggest to smallest?

42   75   342   96
230   18   501   40

_____, _____, _____, _____, _____, _____, _____, _____

NOTE: Your child is learning to write numbers for number words and to order numbers. Ask your child to write the number for fifty-eight.

Complete each number word in any way. Then write the numbers in order from biggest to smallest.

11. forty-_____seven_____ .... _47_

_____thirty_____-six .... _36_

_____-four ..... _____

seventy-_____ ... _____

one	two	three
four	five	six
seven	eight	nine
twenty	thirty	forty
fifty	sixty	seventy
eighty	ninety	hundred

In order: _____,_____, _____, _____

12. _____ hundred fifty-five .... _____

two hundred _____ -two............... _____

four hundred seventy- _____ .... _____

_____ hundred twenty-_____ ...... _____

In order: _____,_____, _____, _____

Challenge

13. What is the biggest number you can make from these digits? What is the smallest? Explain.

_____

_____

_____

Name _____ Date _____

Working with Hundreds, Tens, and Ones

NCTM Standards 1, 2, 6, 8, 9, 10

🔻 TEKS 2.1B, 2.3A, 2.13A

What number is shown by the blocks?

1.

64

2.

3.

4.

5.

6.

7.

8.

 NOTE: Your child is learning that you name a number by combining hundreds with hundreds, tens with tens, and ones with ones.

9. Which pairs make 100? Circle them as fast as you can.

Sums of 100 Search								
40 60	60 40	60 50	30 60	30 70	80 20	10 90	50 60	20 80
80 20	90 10	10 80	90 10	60 30	20 90	20 80	40 60	50 50
80 20	90 20	30 60	20 90	70 30	70 20	60 30	70 30	20 80
40 60	50 60	50 50	30 70	40 60	50 50	70 30	80 20	30 80

10. Pick one of the number pairs that you circled. How do you know that it has a sum of 100?

_____

_____

_____

Problem Solving

11. Marcy is thinking of a number. It has 6 tens, 4 ones, 2 hundreds. What is her number? _____

Show Marcy's number in three different ways.

Chapter 3
Lesson 9

Problem Solving Strategy
Draw a Picture

NCTM Standards 1, 2, 6, 7, 8, 9

TEKS 2.12A, 2.12B, 2.12C

Understand
Plan
Solve
Check

1. Mr. Brown's art room has 1 box of pencils, 2 packages of pencils, and 3 loose pencils. Ms. Gold's art room has 1 box and 5 packages of pencils. How many pencils are there in all?

| 1 package = 10 pencils |
| 1 box = 100 pencils |

_____ pencils

2. There are 4 children in line. Laura is last. Charles is between Kim and Dan. Kim is next to Laura. Who is first in line?

_____ is first.

3. Dogs, people, and fish live in a house. There are 8 heads and 18 legs. There are 3 dogs. How many people and fish are in the house? (Hint: A fish has no legs.)

3

_____ dogs _____ people _____ fish

NOTE: Your child is exploring different ways to solve problems. Sometimes drawing a picture is an efficient way to solve a problem.

Problem Solving Test Prep

1. The drawing shows how many pennies Tara saved each day. If the pattern continues, how many pennies will she save on Thursday?

Monday

Tuesday

Wednesday

   Ⓐ 8   Ⓒ 12

   Ⓑ 10   Ⓓ 16

2. Denise has 5 pages of 10 stickers and 3 loose stickers. Tom has 20 more stickers than Denise. How many stickers does Tom have?

   Ⓐ 35   Ⓒ 59

   Ⓑ 37   Ⓓ 73

✏ Show What You Know

3. Ty asked 9 friends to name their favorite sport. Most like soccer best. The rest like football best. How many friends do you think like each sport?

_____ like soccer best.

_____ like football best.

Tell how you found your answer.

_____

_____

4. Rob has 8 crayons. Lena has 7 more crayons than Rob. What number sentence can you use to find how many crayons Lena has?

_____

Tell how you found your answer.

_____

_____

_____

Chapter 3 Review/Assessment

NCTM Standards 1, 6, 7, 8, 9, 10

1. How many are there? Use the set of 10 to estimate.
Then count to find the total. Lesson 1

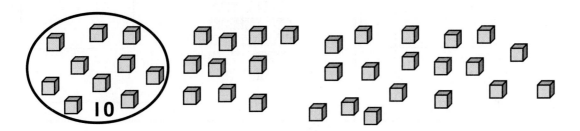

Estimate _____   Count _____

What number do the blocks show? Lesson 2

2.

_____   _____   ⟹   _____
 tens   ones

3.

_____   _____   _____   ⟹   _____
hundred   tens   ones

What is missing? Draw symbols and write numbers. Lessons 3, 4

4.

7 tens 5 ones = _____ + _____ = _____

5.

_____ hundreds _____ tens _____ ones

_____ + _____ + _____ = 309

Show each number with the fewest blocks
Draw symbols for the blocks. Lesson 5

6.

7.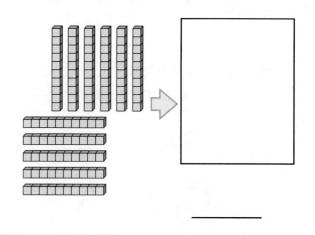

_____

_____

Write the numbers. Then write <, >, or =. Lessons 6, 8

8.

9.

10. Write the numbers in order from biggest to smallest. Lessons 6, 7

150   51   75   501   455   40

_____, _____, _____, _____, _____, _____

Problem Solving Lesson 9

11. Fran, George, Helena, and Jaime are in line for ice cream. Fran is third in line. George is between Jaime and Fran. Who is last in line?

_____

Name _____

Addition and Subtraction with Place Value
Combining Blocks ✏
TEKS 2.1A, 2.3B, 2.13A

Put all of your base-ten blocks in a pile.

STEP 1 Making Sets

Choose any 6 base-ten blocks from the pile.
What blocks did you and your partner choose?

	How many rods?	How many units?	Total Value
Child 1			
Child 2			

STEP 2 Combining Sets

What is the value of the blocks in both sets? _____

Explain how you found the total.

_____

_____

STEP 3 Recording Totals

How can you record the way you combined the
two sets of blocks?

_____

_____

_____

_____

Investigation

Dear Family,

Today we started Chapter 4 of *Think Math!* In this chapter, I will work with base-ten blocks, symbols, and numbers to learn how to add and subtract. I will also work with dollars, dimes, and pennies. There are NOTES on the Lesson Activity Book pages to explain what I am learning every day.

Here are some activities for us to do together at home. These activities will help me practice adding and subtracting.

Love,

_____

Family Fun

Adding Dimes and Pennies

Work with your child to practice addition using dimes and pennies.

- Gather a bunch of dimes and pennies for this activity.

- Each of you takes a handful of pennies and dimes and puts them in front of you.

- Work together to count each handful of coins and write the amounts on a sheet of paper.

43¢ 38¢

- Combine the coins from both handfuls.

- Work together to find how much money there is in all. If you have 10 pennies, trade them for a dime. Try counting the dimes first and then adding the pennies. Write the total.

| 43¢ |
| 38¢ |
| 81¢ |

- Repeat with other handfuls of coins.

Cross Number Puzzles

Work with your child to complete a Cross Number Puzzle.

- A Cross Number Puzzle is like a crossword puzzle for numbers.

- Together, complete this puzzle by adding across and down. For each row, write the sum of the two numbers in the shaded box. Do the same for each column. In the green shaded box, write the sum of the rows and columns. The sum of the columns should always be the same as the sum of the rows!

4	8	
6	2	

4	8	12
6	2	8
10	10	20

Exploring Addition with Base-Ten Blocks

NCTM Standards 1, 2, 6, 7, 8, 9, 10

TEKS 2.1A, 2.1B, 2.3B

☐ is 100, | is 10, and • is 1.

What is the total? Draw the symbols.

1.

In Stock	New	Total

2.

In Stock	New	Total

3.

In Stock	New	Total

4.

In Stock	New	Total

5.

In Stock	New	Total

6.

In Stock	New	Total

 NOTE: Your child is learning to add hundreds, tens, and ones to find the sum. Your child can also record the blocks with symbols.

What is the total? Draw the symbols.

7.

In Stock	New	Total

8.

In Stock	New	Total

9.

In Stock	New	Total

10.

In Stock	New	Total

Problem Solving

11. Jack has 3 rods and 8 units. He wants to trade for a flat. What blocks does he need in order to make the trade? Use words, numbers, or pictures to explain.

_____

_____

_____

_____

_____

Name _____ Date _____

Exploring Subtraction with Base-Ten Blocks

NCTM Standards 1, 2, 6, 8, 9, 10

TEKS 2.1A, 2.3B

What is missing? Draw the symbols.

1.

In Stock	Sold	Left

2.

In Stock	Sold	Left

3.

In Stock	Sold	Left

4.

In Stock	Sold	Left

5.

In Stock	Sold	Left

6.

In Stock	Sold	Left

NOTE: Your child is learning to subtract with hundreds, tens, and ones. Ask your child to explain how to find the answer for Problem 6.

What is missing? Draw the symbols.

7.

In Stock	Sold	Left

8.

In Stock	Sold	Left

9.

In Stock	Sold	Left

10.

In Stock	Sold	Left

11.

In Stock	Sold	Left

Open a package if you need ones.

Challenge

12. What is missing?

In Stock	Sold	Left

Name _____ Date _____

Wonder Wheel Addition and Subtraction

NCTM Standards 1, 2, 6, 8, 9, 10

TEKS 2.1A, 2.1B, 2.3B

What is the total? Draw the symbols.

1.

In Stock	New	Total

2.

	In Stock
	New
	Total

3.

	In Stock
	New
	Total

4.

	In Stock
	New
	Total

NOTE: Your child is learning to add and subtract two- and three-digit numbers. Your child can also record the blocks with symbols.

What is missing? Draw the symbols.

5.

In Stock	New	Total

6.

In Stock	New	Total

7.

	In Stock
	New
	Total

8.

	In Stock
	New
	Total

Challenge

9. Show two different ways to make the same total.
Draw the symbols.

In Stock	New	Total

In Stock	New	Total

Introducing the Cross Number Puzzle

NCTM Standards 1, 2, 6, 7, 8, 9, 10

TEKS 2.3B

What is missing?

Rules:
- Make the same amount on both sides of the thick line.
- You cannot have more than 9 of any symbol in each box.

Work across and down. Follow the rules.

1.

2.

3.

4.

5.

6.

7.

 NOTE: Your child is learning to solve Cross Number Puzzles using symbols. These puzzles will help your child understand how to add and subtract numbers.

What is missing?

8.

9.

10.

11. Make your own.

Challenge

12. What is missing?

How can you check your answers?

_____

_____

_____

_____

Name _____   Date _____

Addition and Subtraction

NCTM Standards 1, 2, 6, 9, 10

TEKS 2.1B, 2.3B

What is missing? Draw the symbols and write the numbers.

I. 36

2.

3.

4.

5. 132

6. 40

7. 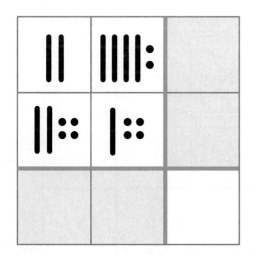 62

NOTE: Your child is learning to add and subtract two- and three-digit numbers using Cross Number Puzzles.

What is missing? Draw the symbols and write the numbers.

8.

9.

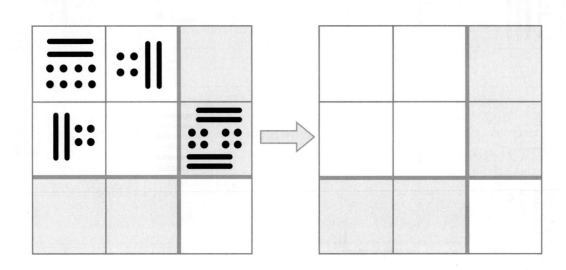

Challenge

10. What is missing?

	□	

→

20	4	
	9	

Chapter 4
Lesson **6**

Adding a Multiple of 10

NCTM Standards 1, 2, 6, 7, 8, 9, 10
TEKS 2.1B, 2.3B

What is missing?

1.

30

4

34

2.

4

3.

64

4.

60

5.

134

6.
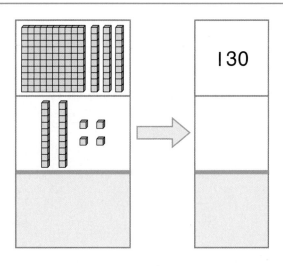

130

NOTE: Your child is learning to add multiples of 10.
Ask your child to solve 20 + 16.

What is missing?

7.

8.

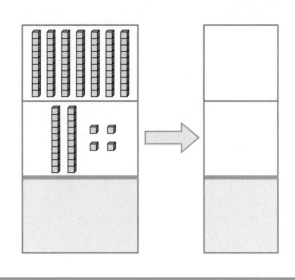

Use the pattern to help you.

9.

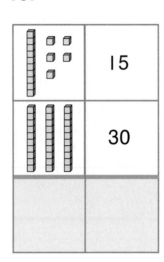

	15
	10

10.

	15
	30

11.

	15
	40

12.

	15
	70

Problem Solving

13. Mrs. Sung buys a roll of 100 stamps.
She has 32 stamps at home. How many
stamps does Mrs. Sung have in all?

Use words, numbers, or pictures to explain.

_____ stamps

Chapter 4
Lesson 7

Fewest Dimes and Pennies

NCTM Standards 1, 2, 6, 7, 8, 9, 10

TEKS 2.14

How can you show 24¢ using only dimes and pennies?

| 24¢ ||
Dimes	Pennies
0	24
1	14
2	4

This way uses the fewest coins.

How can you show each money amount with the fewest coins? Use only dimes and pennies.

Money Amount	How many dimes?	How many pennies?	How much money?
1. twenty-four cents	2	4	24 ¢
2. twenty-five cents			¢
3. twenty-six cents			¢
4. thirty-six cents			¢
5. forty-six cents			¢
6. fifty-two cents			¢
7. seventeen cents			¢
8. ten cents			¢

NOTE: Your child is learning to represent money amounts using the fewest dimes and pennies. Together, show 43¢ using the fewest dimes and pennies.

Show each money amount as many ways as you can.
Use only dimes and pennies.

9.

16¢	
Dimes	**Pennies**
0	16

How many coins are used to show 16¢ with the fewest dimes and pennies?

_____ coins

10.

37¢	
Dimes	**Pennies**

How many coins are used to show 37¢ with the fewest dimes and pennies?

_____ coins

11. Max has 42¢ in dimes and pennies. He has the fewest coins for that amount. How many coins does he have? Tell how you know.

_____ coins

_____

_____

_____

Problem Solving

12. Tara has only pennies and dimes in her pocket. She takes out 5 coins. What is the most money that Tara could have? Explain.

_____ ¢

_____

_____

Name _____ Date _____

Fewest Dollars, Dimes, and Pennies

NCTM Standards 1, 2, 6, 7, 8, 9, 10
TEKS 2.3D, 2.14

How can you show $2.43 with the fewest bills, dimes, and pennies?

Use as many dollar bills as you can. Use as many dimes as you can.

2 dollars + 4 dimes + 3 pennies = $2.43

How can you show each money amount with the fewest bills and coins? Use only dollars, dimes, and pennies.

	Money Amount	How many dollars?	How many dimes?	How many pennies?
1.	$1.62	1	6	2
2.	$1.57			
3.	$1.93			
4.	$2.14			
5.	$2.30			
6.	$3.41			
7.	$3.08			

 NOTE: Your child is learning to represent money amounts with the fewest dollars, dimes, and pennies. Together, show $1.23 using 1 bill, 2 dimes, and 3 pennies.

How much money is there?

P = = 1¢ D = = 10¢

8.

D D P	21¢
D P P	12¢
D D P D P P	___ ¢

9.

D P P P P P	___ ¢
D P P P P P	___ ¢
D D D	___ ¢

 10. Joanne has $2.63. What is the fewest bills and coins she could have? Explain how you know.

_____

_____

_____

Problem Solving

11. Jared has 1 dollar bill, 4 dimes, and 16 pennies. How much money does he have?

$ _____

Can he show the amount with fewer dollars, dimes, and pennies? Use words, numbers, or pictures to explain.

Problem Solving Strategy
Work Backward

NCTM Standards 1, 2, 6, 7, 8, 9, 10
TEKS 2.12B, 2.12C

Understand
Plan
Solve
Check

1. Brad's basketball team scored 14 points in the second half. They had 56 points at the end of the game. How many points did the team get in the first half?

_____ points

Explain how you found your answer. _____

_____

_____

2. Jen reads every night for 20 minutes. Tonight she finishes at 8:00. What time did she start reading?

____:____

Explain how you found your answer. _____

_____

_____

3. The bus stops and 8 people get on and 10 people get off. Now there are 19 people on the bus. How many people were on the bus before it stopped?

_____ people

Explain how you found your answer. _____

_____

_____

NOTE: Your child is exploring different ways to solve problems. Sometimes *working backward* is an efficient way to solve a problem

Problem Solving Test Prep

1. Juan puts 5¢ a day in his bank. How much will he save in 6 days?

   Ⓐ 5¢

   Ⓑ 15¢

   Ⓒ 30¢

   Ⓓ 35¢

2. Mary took 3 pencils out of a new box. There are 15 pencils left in the box. How many pencils were in the box to start?

   Ⓐ 3 pencils

   Ⓑ 9 pencils

   Ⓒ 12 pencils

   Ⓓ 18 pencils

Show What You Know

3. Ann, Tim, and Carlos are in a race. How can they finish if a girl comes in first?

   _____   _____   _____
     first   second   third

   Explain your answer.

   _____

   _____

   _____

   _____

   _____

4. Lee has only pennies and dimes. He has 6 coins. What is the most money he can have?

   _____¢

   Explain your answer.

   _____

   _____

   _____

   _____

What is missing? Draw the symbols. Lessons 1–4

1.

In Stock	New	Total

2.

In Stock	Sold	Left

3.

In Stock	New	Total

4.

	In Stock
	New
	Total

5. What is missing? Draw the symbols and the numbers. Lesson 5

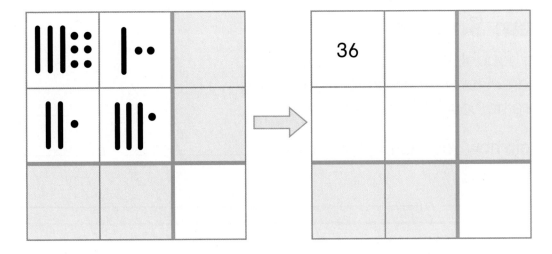

What is missing? Lesson 6

6.

7.
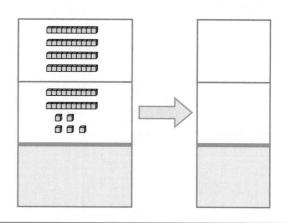

How can you show each amount with the fewest coins? Lesson 7

	Money Amount	How many dimes?	How many pennies?	How much money?
8.	thirty-six cents			¢
9.	ninety-seven cents			¢

How much money is there? Lesson 8

10.

11.
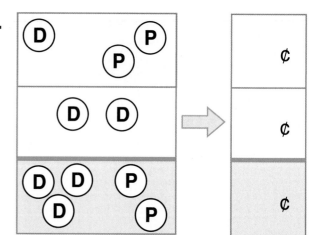

Problem Solving Lesson 9

12. Greg bought 12 toy cars. Now Greg has 28 cars. How many cars did Greg have before he bought these new ones?

_____ cars

Explain how you found your answer. _____

_____

Probability and Data
Probability of Pulling Pairs ✏️

You need
- two pairs of cubes in different colors
- paper lunch bag

Name _____

What is the chance you will pull out two cubes that are the same color?

STEP 1 Pulling Out Two

Pull out two cubes. Are both cubes the same color?
Use a ✔ to record each turn.

Matching Pair	
Yes	**No**

Will you always pull out a matching pair? Explain. _____

_____

STEP 2 Pulling Out Three

Pull out three cubes. Are two of the cubes the same color? Record each turn.

Matching Pair	
Yes	**No**

Tell about your results. _____

Why do you think this happens? _____

_____

_____

STEP 3 Pulling Out Four

Pull out four cubes. Will you get a matching pair? Explain.

_____

_____

Investigation

Dear Family,

Today we started Chapter 5 of *Think Math!* In this chapter, I will explore probability, picture graphs, real-object graphs, bar graphs, pictographs, and even line graphs. There are NOTES on the Lesson Activity Book pages to explain what I am learning every day.

Here are some activities for us to do together at home. These activities will help me understand probability and data.

Love,

_____

Family Fun

What Are You Pulling?

Work with your child to play a probability game called *What Are You Pulling?*

■ Put 30 of the same food item into a paper lunch bag. You might use candies, cereal pieces, or crackers in different colors or shapes. Do not let your child see what is inside the bag.

■ Take turns with your child. Pull out one item at a time from the bag without looking inside. Then replace the item and mix them up. Record the results of each turn. After taking 10 turns, start asking this question: What do you think you will pull out now?

■ With each turn, your child should get better at making predictions. Look back at what has already been pulled out before making a prediction.

■ Continue for as long as you wish. Then, if you want, eat the items together!

Word Search

Work with your child to circle these words in the puzzle.

BAR GRAPH   DATA
IMPOSSIBLE   KEY
PICTOGRAPH   LIKELY
PICTURE GRAPH   POSSIBLE
LINE GRAPH   PREDICT
SYMBOL   TALLY

P	E	S	Y	L	L	D	Y	I	M	T
I	B	A	R	G	R	A	P	H	L	A
C	M	O	E	S	H	T	O	E	I	L
T	A	L	L	Y	H	A	S	R	K	I
U	L	A	K	E	Y	P	S	A	E	N
R	I	M	P	O	S	S	I	B	L	E
E	N	A	P	H	G	E	B	B	Y	G
G	O	S	Y	M	B	O	L	A	N	R
R	A	P	H	A	G	R	E	S	I	A
A	P	R	E	D	I	C	T	O	T	P
P	I	C	T	O	G	R	A	P	H	H
H	M	T	U	K	L	I	T	R	U	N

Exploring Probability

NCTM Standards 1, 2, 5, 6, 7, 8, 9, 10
TEKS 2.11C, 2.13A, 2.13B, 2.14

There are 10 cubes in each bag. Some are red and some are blue. Write how many of each and color the cubes to match the story.

1. Joe is more likely to pull out a red cube than a blue cube from the bag.

   _____ red   _____ blue

2. Shamari is equally likely to pull out a red cube or a blue cube.

   _____ red   _____ blue

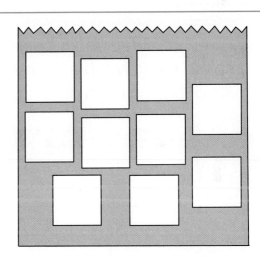

3. Make your own story.

   Matt is _____ likely

   to pull out a _____ cube

   than a _____ cube.

   _____ red   _____ blue

NOTE: Your child is learning about the likelihood of events. Talk about events that are more likely and less likely to happen in your everyday lives.

Each bag has different cubes. Sue pulls out a cube and records the color. She does this 20 times. She puts the cube back each time. Which color do you think she is more likely to pull out next?

4.

Red	Blue
~~IIII~~ ~~IIII~~ ~~IIII~~ III	II

_____

5.

Red	Blue
~~IIII~~	~~IIII~~ ~~IIII~~ ~~IIII~~

_____

6.

Red	Blue
	~~IIII~~ ~~IIII~~ ~~IIII~~ ~~IIII~~

_____

7. This bag has 5 red and 5 blue cubes. Complete the tally table to show possible results of pulling out a cube 20 times.

Red	Blue

Problem Solving

8. Sue pulled out a cube from a bag 20 times. She made this tally table. What are the chances that a green cube was in this bag? Explain.

Red	Blue
~~IIII~~ I	~~IIII~~ ~~IIII~~ IIII

_____

_____

_____

Using Real-Object Graphs and Picture Graphs

NCTM Standards 1, 2, 4, 5, 6, 7, 8, 9, 10

TEKS 2.11B

Each child in Room 2 has a penny. The children lined up to show the years their pennies were minted. They recorded the data like this.

The mint year is the year a coin was made.

Mint Years for Our Pennies

1. The oldest penny was minted in _____ .

2. _____ has the newest penny.

3. There are _____ children in Room 2.

4. Melanie's penny was minted in _____ .

5. Write your own sentence about the picture graph.

_____

  NOTE: Your child is learning to make and use real-object graphs and picture graphs to show information.

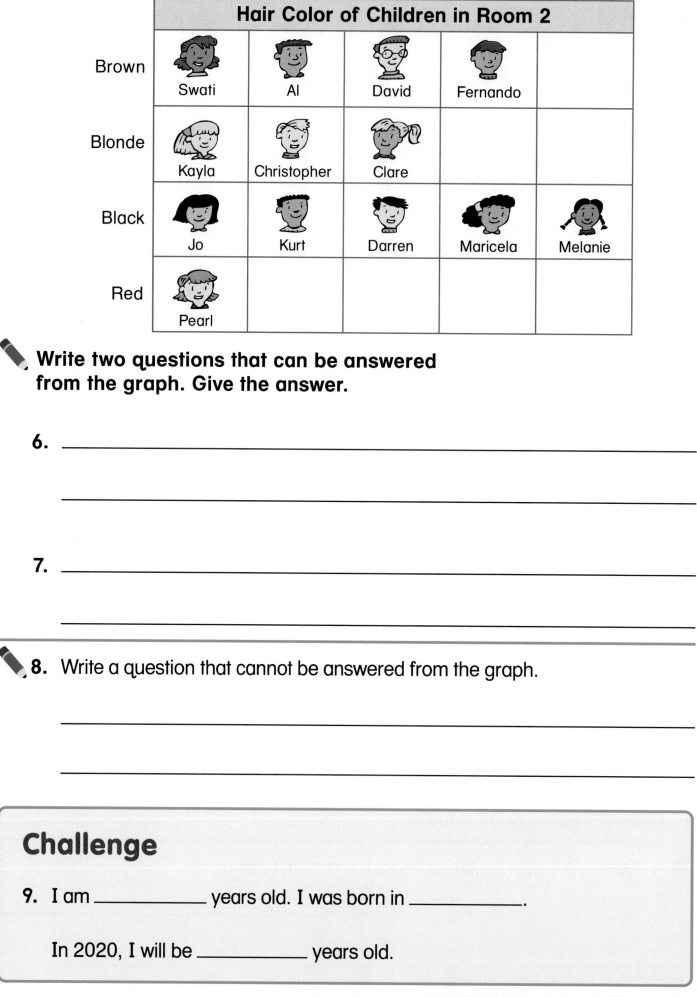

Hair Color of Children in Room 2				
Brown	Swati	Al	David	Fernando	
Blonde	Kayla	Christopher	Clare		
Black	Jo	Kurt	Darren	Maricela	Melanie
Red	Pearl				

✏ **Write two questions that can be answered from the graph. Give the answer.**

6. _____

_____

7. _____

_____

✏ **8.** Write a question that cannot be answered from the graph.

_____

_____

Challenge

9. I am _____ years old. I was born in _____.

In 2020, I will be _____ years old.

Chapter 5
Lesson 3

Using Bar Graphs to Investigate Probability

NCTM Standards 1, 2, 4, 5, 6, 7, 8, 9, 10
TEKS 2.11A, 2.11B, 2.11C

You have two number cubes. One cube has numbers 1 to 6. The other has numbers 7 to 12.

1. What are all of the possible addition facts you could make by tossing the number cubes?

+	7	8	9	10	11	12
1	1 + 7					
2		2 + 8				
3			3 + 9			
4						
5						
6						

2. What are all of the possible sums?

_____

3. How many different ways can you make each sum by tossing the number cubes? Tally to record.

Sum	8	9	10	11	12	13	14	15	16	17	18
Tally	I	II									
Number of Ways	1	2									

NOTE: Your child is learning about bar graphs and is using this representation to record the results of an addition and probability game.

Play the game with a classmate. Toss two number cubes. Color a box in the graph for each sum. Which numbers win? Answer Questions 4 and 5 as you play. Answer Question 6 at the end.

Number Race Gameboard

Number of Tosses (y-axis: 0, 1, 2, 3, 4, 5, 6, 7, 8, 9, 10)

Sum (x-axis: 7, 8, 9, 10, 11, 12, 13, 14, 15, 16, 17, 18)

4. Which number got to the blue line first? _____

5. Which number got to the green line first? _____

6. Which number finished the race first? _____

Challenge

7. How could you change the game so that one sum is certain to win every time?

_____

_____

_____

Chapter 5
Lesson 4

Making and Using Bar Graphs

NCTM Standards 1, 2, 4, 5, 6, 7, 8, 9, 10

⬇ TEKS 2.11A, 2.11B

This class list gives the names for all children in Room 2.

Check off each name as you record it in the graph.

✔ Melanie	Christopher	Kayla	Clare
Jo	Kurt	Swati	Maricela
Fernando	Darren	Pearl	Al
David			

I. Use the class list. Make a bar graph showing the lengths of the names.

Length of Names in Room 2

Number of Names

6
5
4
3
2
1
0

1 2 3 4 5 6 7 8 9 10 11 12

Number of Letters in the Name

2. How many letters are in the longest name? _____ letters

3. How many names have 5 letters? _____ names

🏠 **NOTE:** Your child is learning how to create a bar graph. Your child is also learning to identify whether a question can be answered by a table or a graph.

4. Use the table to make a bar graph. Remember to label the graph.

Length of Names in Room 3	
Number of Letters	**Name**
3	Ned
4	Anna John Ling
5	Kayla Nikil
7	Chelsea Tabitha
8	Benjamin
9	Elizabeth

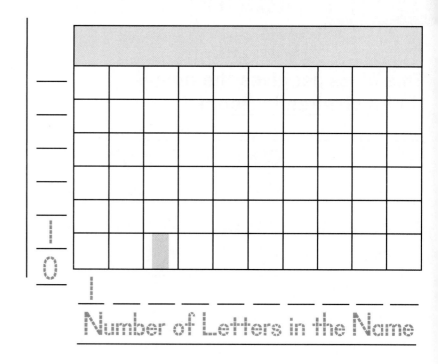

Number of Letters in the Name

Answer each question. Circle whether you use the table or graph to find the answer.

5. How many letters are in
the names for the most children? _____ letters   table   (graph)

---

6. What is the shortest name?

_____   table   graph

---

7. How many more names have
4 letters than 9 letters?   _____ names   table   graph

Challenge

8. What can you learn from the table that is not in the graph?

_____

_____

Making and Using Pictographs

NCTM Standards 1, 2, 5, 6, 7, 8, 9, 10
TEKS 2.11A, 2.11B, 2.14

Tom asked the children in each grade at school whether they have a dog. Then he made this pictograph.

Children at Bayles School with Dogs					
			☺		
			☺		
		☺	☺		☺
☺		☺	☺		☺
☺		☺	☺		☺
☺	☺	☺	☺	☺	☺
1st	2nd	3rd	4th	5th	6th
		Grade			

Key: Each ☺ stands for 5 children.

 1. How many first graders have dogs? _____ children

How does the graph show that? _____

_____

2. How many more fourth graders
than third graders have dogs?   _____ children

How did you figure that out? _____

_____

NOTE: Your child is learning how to create a pictograph
where a symbol stands for more than one piece of data.

3. Use the tally table to make a pictograph. Remember to choose a symbol and make a key.

Children at Bayles with Cats	
Grade	**Tally**
1	ⵜⵜ ⵜⵜ IIII
2	ⵜⵜ ⵜⵜ ⵜⵜ I
3	ⵜⵜ III
4	II
5	ⵜⵜ ⵜⵜ ⵜⵜ III
6	ⵜⵜ I

Children at Bayles School with Cats

Grade									
1st									
2nd									
3rd									
4th									
5th									
6th									

Key: Each _____ stands for _____ children.

4. Write your own sentence about the pictograph.

_____

_____

Write two questions that can be answered from the graph.

5. _____

6. _____

Problem Solving

7. Three children in Kindergarten have cats. How could you show this in the graph if each symbol stands for 2 children?

_____

_____

Name _____ Date _____

Graphing Change Over Time

NCTM Standards 1, 2, 4, 5, 6, 7, 8, 9, 10

TEKS 2.13A, 2.14

This table shows typical snowfall for Chicago, Illinois. The amount of snow for March has been left out.

Snowfall in Chicago							
Month	Oct	Nov	Dec	Jan	Feb	Mar	Apr
Inches of Snow	$\frac{1}{2}$	2	8	11	8	?	2

1. Make a line graph of the data in the table.

Find the point on the graph where the month and the inches intersect.

 2. Use the graph to estimate how much snow fell in March. Explain your answer.

_____

_____

NOTE: Your child is learning how to create and interpret line graphs to show change over time.

Draw a line from each story to the graph that matches it best.

3. Lakisha saved money in her piggy bank. She sometimes spent some of it.

4. The teacher in Room 2 recorded how many children came to school each day.

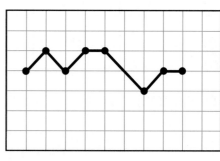

5. Al completed the same addition table several times. He recorded how long it took each time.

6. How are a line graph and a bar graph different?

_____

_____

Problem Solving

7. Write a story about this line graph. Think about what makes sense for a line that keeps going up.

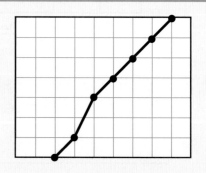

_____

_____

Name _____ Date _____

Problem Solving Strategy
Make a Table ✏

NCTM Standards 1, 2, 4, 5, 6, 7, 8, 9, 10
TEKS 2.12B, 2.12C

Understand
Plan
Solve
Check

1. Nev spins both spinners shown.
   What sum is she most likely to get?

   _____

Spinner A **Spinner B**

Spinner A	1	1	1	3	3	3	4	4	4
Spinner B	4	5	7	4	5	7			
Sum	5								

2. How many tires will Seth need for 7 model cars?

Cars	1	2	3				
Tires							

_____ tires

3. Cindy has 3 coins in her pocket.
   She only has coins worth 5¢ or less.
   How much money might she have?

Pennies	Nickels	Total Money
0	3	

NOTE: Your child is exploring different ways to solve problems. Sometimes making a table is an efficient way to solve a problem.

Problem Solving Test Prep

1. Kip has 12 blocks and 2 plates. He puts half of the blocks on each plate. How many blocks are on each plate?

   Ⓐ 24 blocks

   Ⓑ 12 blocks

   Ⓒ 6 blocks

   Ⓓ 3 blocks

2. Sara had 25¢ in her bank. She put some more coins into the bank. Then she had 38¢. How much money did she put in the bank?

   Ⓐ 63¢

   Ⓑ 13¢

   Ⓒ 5¢

   Ⓓ 3¢

Show What You Know

3. Ivana has 32 marbles in a solid color and 15 that are striped. How many marbles does she have in all?

   _____ marbles

   Explain how you found the answer.

   _____

   _____

   _____

   _____

   _____

4. Carlos drove 183 miles on Saturday. He drove farther on Sunday. The number of miles has the same digits in a different order. How far did Carlos drive on Sunday?

   _____ miles

   Explain how you found the answer.

   _____

   _____

   _____

   _____

Review/Assessment
NCTM Standards 1, 2, 4, 5, 6, 7, 8, 9, 10

1. Which color is more likely to be pulled out
next from the bag? Use the tally table to help. Lesson 1

Red	Blue
卌 卌 卌 卌	

_____

Use the picture graph to complete each sentence. Lesson 2

2. There are _____ children
at the playground.

3. _____ is on the climber.

4. There are _____ more children
on the slide than on the climber.

Children at the Playground

swings — Ted, Kim, David, Deena

slide — Noah, Lily, Zoe

climber — Ben

Use the bar graph to answer each question. Lessons 3, 4

Length of Names in Our Class

Number of Names (y-axis: 0–6)
Number of Letters in the Name (x-axis: 1–12)

5. How many letters are in the longest name? _____ letters

6. How many names are 4 letters long? _____ names

7. Use the tally table to make a pictograph.
Choose a symbol and make a key. Lesson 5

Crayons in the Box	
Color	**Tally**
blue	‖‖ ‖‖ ‖
green	‖‖ ‖‖ ‖‖ ‖
red	‖‖ ‖‖ ‖
yellow	‖‖ ‖‖

Crayons in the Box								
blue								
green								
red								
yellow								

Key: Each _____ **stands for** _____ **crayons.**

Answer each question about the line graph. Lesson 6

8. How many hot dogs were sold at 3:00?

_____ hot dogs

9. Did hot dog sales go up or down from 4:00 to 6:00?

_____

Hot Dog Sales at the Carnival

Number of Hot Dogs Sold / Time

Problem Solving Lesson 7

10. Sal spins both spinners shown. How many different sums can he spin?

_____ sums

Spinner A   Spinner B

Spinner A	I	I	3	3
Spinner B	I	2	I	
Sum				

Chapter 6

Measuring Time
School Schedule ✏

Discuss the activities you do in school.

STEP 1 Describing Activities

What activities happen every day in school?

_____

_____

_____

STEP 2 Ordering Activities

Draw each activity on a card. In what order
do you do the activities in school? Put the cards
in order starting with the first activity of the day.

STEP 3 Thinking About Time

Which activities seem to take a long time?

_____

_____

Which activities seem to take a short time?

_____

_____

Investigation

Dear Family,

Today we started Chapter 6 of *Think Math!* In this chapter, I will learn to read time. I will also solve problems involving time. There are NOTES on the Lesson Activity Book pages to explain what I am learning every day.

Here are some activities for us to do together at home. These activities will help me understand concepts related to time.

Love,

_____

Family Fun

How Much in 5 Minutes?

Have your child complete activities that take 5 minutes.

- Ask your child, "What do you think you could do in 5 minutes?" Make a list. Possibilities include: make my bed; write the alphabet; and practice my math facts.

- Ask your child to choose one of the activities on the list to do as you time it. How long did it take?

- Repeat the activity for several days in a row. Together, graph the amount of time the activity takes every day.

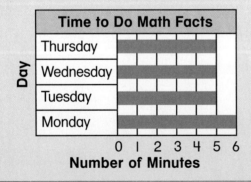

Time to Do Math Facts

Day	Number of Minutes
Thursday	
Wednesday	
Tuesday	
Monday	

0 1 2 3 4 5 6

What Time Is It?

Work with your child to practice telling time.

- Discuss with your child the activities your family does every day. For example, get up in the morning, get washed and dressed, eat meals, come home, go to bed, and so on.

- Whenever a new activity begins or ends, ask your child, "What time is it?" Have your child read a clock or a watch to tell the time. Repeat this activity throughout the day.

It's 7:00. Time to wake up!

Chapter 6
Lesson 1

Exploring Time

NCTM Standards 4, 5, 6, 7, 8, 9, 10

🖐 TEKS 2.10C, 2.13A, 2.14

1. How many smiley faces can you draw in 10 seconds?
Have a partner time 10 seconds as you draw.

How many smiley faces did you draw? _____ faces

2. How many smiley faces can you draw in 1 minute?
Have a partner time 1 minute as you draw.

How many smiley faces did you draw? _____ faces

 NOTE: Your child is learning different
units of time including seconds, minutes,
and hours.

3. Draw a picture of something you do that takes about 1 hour.

4. Does an hour seem long or short to you? Tell why.

_____

_____

_____

Challenge
Circle the better estimate for each time.

5. About how long does it take to say the alphabet?

1 second   1 minute

6. About how long does it take to play a game of soccer?

1 minute   1 hour

Chapter 6
Lesson 2

Minutes in an Hour

NCTM Standards 4, 6, 8, 10

TEKS 2.10B, 2.12A, 2.12B, 2.12C, 2.13A, 2.14

What time is it?

1.

5:00

2.

:

3.

:

4.

:

5.

:

6.

:

7. Show any time to the half hour. Draw
   the clock hands and write the time.

:

NOTE: Your child is learning to tell time to
the hour and half hour. Ask your child to read
the clock when the time is on the hour.

What is missing?

8.

I hour later

Draw the clock hands.

3:30

9.

2 hours later

Problem Solving

10. A machine makes 1 toy every minute. How many toys can the machine make in 25 minutes? _____ toys

11. A machine makes 2 boxes every minute. How many boxes can the machine make in 1 hour? Explain your answer. _____ boxes

_____

_____

_____

Chapter 6
Lesson 3

Telling Time to 10 Minutes

NCTM Standards 3, 4, 6, 7, 8, 9, 10

TEKS 2.10B, 2.12A, 2.12B, 2.13A, 2.14

What time is it?

1.

3 :

2.

:

3.

:

4.

_20_ minutes after _8_

5.

_____ minutes after _____

6.

_____ minutes before _____

7.

_____ minutes before _____

 NOTE: Your child is learning to tell time to 10 minutes and describe the time as before and after the hour.

8. Tina travels 1 mile every 10 minutes. What time will she get to each mile flag? Draw the clock hands.

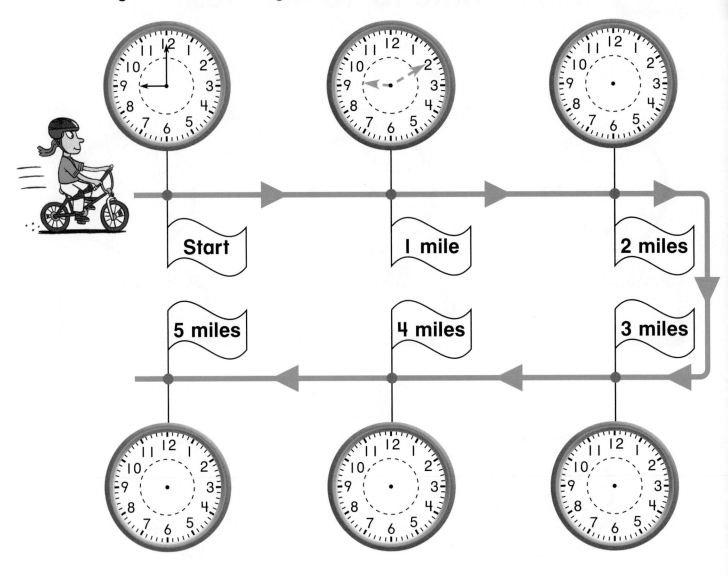

Start 1 mile 2 miles

5 miles 4 miles 3 miles

Problem Solving

9. Dana starts her homework at 3:10. It takes 20 minutes to finish. What time is she done? _____ : _____

Use words, numbers, or pictures to explain.

Chapter 6
Lesson 4

How Far? How Fast?

NCTM Standards 3, 4, 6, 8, 9, 10
TEKS 2.10B, 2.12A, 2.12B, 2.13A, 2.14

What is missing?

1.

 30 minutes later

2.

 half an hour later

3.

 I hour later

4.

 I and a half hours later

5. Make your own.

 2 and a half hours later

NOTE: Your child is learning to read and write time. Ask your child what time it will be in I hour.

6. Nick walks 2 miles in 1 hour. He starts at 4:00. What time will he get to each mile flag?

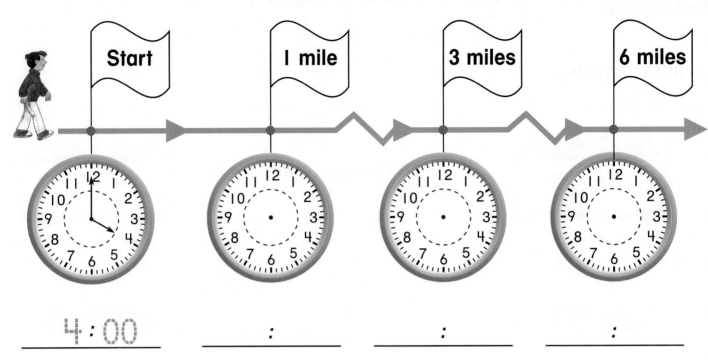

| Start | 1 mile | 3 miles | 6 miles |

4:00 _____   _____ : _____   _____ : _____   _____ : _____

---

7. How can you write 8:15 in words? Write it in two different ways.

_____

_____

Problem Solving

8. A movie starts at 6:50. It is 2 and a half hours long. What time will the movie end? _____ : _____

Use words, numbers, or pictures to explain.

Telling Time to 5 Minutes

NCTM Standards 4, 8, 9, 10
TEKS 2.10B

Exactly what time is it?

1.

3:05

2.

:

3.

:

4.

:

5.

:

6.

:

About what time is it? Circle the closer time.

7.

5:00   5:30

8.

1:30   2:00

9.

7:30   8:00

NOTE: Your child is learning to tell time to 5 minutes and to estimate time. At a time right after the hour, ask your child, "About what time is it?"

What is missing?

10.

5 minutes later ➡

11.

15 minutes later ➡

Challenge

12. What is missing?

5 minutes earlier			2:15	4:05	
Time	9:30	11:15	2:20		
15 minutes later		11:30			5:15

Chapter 6
Lesson 6

Telling Time to the Minute

NCTM Standards 1, 4, 6, 8, 9, 10
TEKS 2.10B, 2.13A, 2.14

What time is it?

1.

2.

Draw the clock hands.

3.

12:23

4.

10:44

5.

3:54

6.

7:38

 NOTE: Your child is learning to tell time to the minute. Ask your child to tell the exact time several times during the day.

What landmark on the clock can help you find the time? Write the landmark and the exact time.

7.

landmark time _8:30_

exact time _8:28_

8.

landmark time ___:___

exact time ___:___

9.

landmark time ___:___

exact time ___:___

10.

landmark time ___:___

exact time ___:___

11. In Problem 9, how did you use the landmark to find the exact time? _____

_____

_____

Challenge

12. What time is it? How do you know? ___:___

_____

_____

Chapter 6
Lesson 7

Calendar and Ordinal Numbers

NCTM Standards 1, 4, 6, 8, 9, 10
🔸 TEKS 2.13A, 2.14

1. Write the name of the month and the year.
Write the dates in the calender.

_____ , _____

Sunday	Monday	Tuesday	Wednesday	Thursday	Friday	Saturday

2. How many days are in this month?   _____ days

3. What day is the first of the month?   _____

4. What day is the nineteenth of the month? _____

5. What is the date of the third Monday?   _____

🏠 **NOTE:** Your child is learning to read a calendar and work
with ordinal numbers. Together, talk about the dates of
holidays and events this month.

What floor is it?

6. Draw an X on the box for the tenth floor.

7. Color the box for the fifth floor green.

8. Color the box for the twelfth floor blue.

9. Write the number in the top floor.

> The boxes on the side are elevator floors.

10. Chloe got on the elevator at the first floor. The elevator went up 7 floors. On what floor was Chloe then?

_____ floor

11. Brad got on the elevator at the twelfth floor. The elevator went down 6 floors. On what floor was Brad then?

_____ floor

12. Write and solve your own elevator problem.

_____

_____

_____

X
1st floor

Challenge

13. Diego got in the elevator on the first floor. The elevator went up 8 floors. Then it went down 3 floors and then up 5 floors. On what floor was Diego then?

_____ floor

Chapter 6
Lesson 8

Problem Solving Strategy
Look for a Pattern

NCTM Standards 1, 2, 3, 4, 6, 7, 8, 9, 10
TEKS 2.6A, 2.6B, 2.12A, 2.12B, 2.12C

Understand
Plan
Solve
Check

1. A rocket travels 10 miles in 2 minutes.
The rocket lifts off at 7:00. How far has
the rocket traveled by 7:20? _____ miles

Explain how you found your answer. _____

_____

_____

_____

2. The first four soccer shirts are numbered
1, 3, 5, and 7. If this pattern continues, what
would be the number on the eighth shirt? _____

Explain. _____

_____

_____

3. A snail travels 1 foot every 5 minutes.
He starts crawling at 9:30. What time
will it be when the snail travels 5 feet?   :   _____

Explain. _____

_____

_____

NOTE: Your child is exploring different ways to solve
problems. Sometimes using the strategy, *look for a pattern,*
is an efficient way to solve a problem.

Problem Solving Test Prep

1. Sean finished baseball practice at 6:00. The practice was one and a half hours long. What time did practice start?

   Ⓐ 4:00

   Ⓑ 4:30

   Ⓒ 5:30

   Ⓓ 7:30

2. Faith made 18 muffins. Ella made 3 less muffins than Faith. How many muffins did they make altogether?

   Ⓐ 15 muffins

   Ⓑ 21 muffins

   Ⓒ 30 muffins

   Ⓓ 33 muffins

✏ Show What You Know

3. Each tricycle has 3 wheels. How many wheels are there on 6 tricycles?

tricycles	1					
wheels	3					

   _____ wheels

   Explain how you found the answer.

   _____

   _____

   _____

   _____

4. There are 4 children in line. Sam is between Clare and Liddy. Ben is last. Liddy is next to Ben. Who is first in line?

   _____

   Explain your answer.

   _____

   _____

   _____

   _____

Review/Assessment

NCTM Standards 1, 2, 4, 6, 7, 8, 9, 10

1. Draw a picture of
something you do that
takes about 1 minute.
Lesson 1

What time is it? Lessons 2, 3

2.

3.

4.

5.

_____ minutes before _____

What is missing? Lesson 4

6.

30 minutes
later

About what time is it? Circle the closer time. Lessons 5, 6

7.

11:30

12:00

8.

3:45

4:00

Exactly what time is it? Lessons 5, 6

9.

_____ : _____

10.

_____ : _____

Solve. Lesson 7

11. Lou got on the elevator at the eleventh floor. The elevator went down 6 floors. On what floor was Lou then?

_____ floor

Problem Solving Lesson 8

12. A bus travels 1 mile in 1 minute. The bus leaves the station at 6:15. How far has the bus traveled by 6:25?

_____ miles

Explain how you found your answer. _____

_____

_____

_____

Chapter 7

Doubling, Halving, and Fractions
Sharing with a Group ✏️
➡️ TEKS 2.4B

How can you share your snack with others?

STEP 1 Observing

Look at your crackers. Are there enough for you and your partner to share? How do you know?

_____

_____

_____

STEP 2 Sharing

Now work with another partner team. How can you share the new group's crackers equally?

_____

_____

_____

STEP 3 Getting One More

Add one more cracker. Now can you share equally? What can you do?

_____

_____

Investigation

Dear Family,

Today we started Chapter 7 of *Think Math!* In this chapter, I will find double and half of a number. I will also identify and write fractions of an object and a set of objects. There are NOTES on the Lesson Activity Book pages to explain what I am learning every day.

Here are some activities for us to do together at home. These activities will help me understand doubles, halves, and fractions.

Love,

_____

Family Fun

Share the Granola Bar!

Work with your child to play this game. Your child will play a similar game in Lesson 7.

- Prepare a gameboard like this. You also need two different-color pencils and a number cube.

- Take turns with your child. For a turn, toss the number cube. This is the number of people who share the granola bar. Find the bar divided into that many equal pieces, write the fraction in one piece, and color that piece. This shows that you ate it. For example, if you roll a 6, write $\frac{1}{6}$ in one piece of the bar for sixths, and then color it.

- If there are no uncolored pieces left for a fraction, you lose a turn.

- The winner is the last to eat a piece of granola!

Double Your Money

Work with your child to practice doubling a money amount.

- Write a money amount less than 50¢ on a slip of paper. Together, show it with coins.

- Ask your child to double that amount of money using any method. Your child can figure out the amount on a scrap of paper and then show it with coins or figure it out just using the coins.

30¢

- As a challenge, you might ask your child to find half of the original amount.

Exploring One Half

NCTM Standards 1, 6, 8, 9, 10
TEKS 2.2A, 2.13A, 2.13B, 2.14

Circle the pictures that show one half.

1.

2.

3.

4.

5.

6.

7.

8.

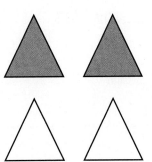

NOTE: Your child is learning to find one half of an object or set of objects.

Color $\frac{1}{2}$ of each picture.

9.

10.

11.

12.

13. Color $\frac{1}{2}$ of the set of circles.
Tell how you know it is $\frac{1}{2}$.

_____

_____

_____

Challenge

14. Dwayne plays video games for half an hour.
How many minutes is that? Explain.

_____ minutes

_____

_____

Finding Half: Even or Odd

NCTM Standards 1, 2, 6, 7, 8, 9, 10
TEKS 2.4B, 2.13A, 2.14

How much is half?

1.

Half of 26 is ___13___.

2.

Half of 14 is _____.

3.

Half of 42 is _____.

4.

Half of 58 is _____.

	Whole	Half
5.	6	3
6.	8	
7.	10	
8.	18	
9.	24	
10.	36	
11.	70	

NOTE: Your child is learning to find half of even and odd numbers. Together, find half of 20.

What number will solve each riddle?

12. I am an even number. I am half of 44. What number am I?

22

13. I am an even number. Half of me is 14. What number am I?

_____

14. I am an even number. Half of me is 36. What number am I?

_____

15. I am an odd number. I am half of 50. What number am I?

_____

16. I am an odd number. Half of me is $11\frac{1}{2}$. What number am I?

_____

17. I am an odd number. Half of me is $9\frac{1}{2}$. What number am I?

_____

18. Make up your own riddle. Ask a classmate to solve it.

_____

_____

_____

Challenge

Write the same number in frames that are the same.

19. ◯ + ◯ = 86

20. ▢ + ▢ = 56

21. ⬡ + ⬡ = 9

22. △ + △ = 25

Name _____   Date/Time _____

Doubling Numbers

NCTM Standards 1, 2, 6, 7, 8, 9, 10
TEKS 2.3A, 2.12A, 2.12B

What is the double of each number?
Draw symbols if you want.

1.

7 doubled is ____14____.

2.

13 doubled is _____.

3.

18 doubled is _____.

4.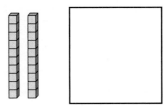

20 doubled is _____.

5.

25 doubled is _____.

6.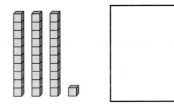

31 doubled is _____.

7.

42 doubled is _____.

8.

56 doubled is _____.

NOTE: Your child is learning
to double numbers. Together,
find the double of 9.

What are the missing numbers? Write each rule.

9.

0	0
1	2
2	4
3	
4	
	10
6	

10.

5	9
6	11
20	39
3	
10	
11	
	29

Rule: _____   Rule: _____

Use doubles to solve.

11. 6 + 6 = _____   6 + 5 = _____   6 + 7 = _____

12. 15 + 15 = _____   15 + 14 = _____   16 + 14 = _____

13. 50 + 50 = _____   49 + 49 = _____   49 + 51 = _____

14. 20 + 20 = _____   19 + 19 = _____   17 + 19 = _____

Problem Solving

15. Kyle has a recipe that makes 12 cups of punch.
Kyle doubles the recipe. How many cups
will he make?   _____ cups

Kyle doubles the recipe again. How many
cups will he have now?   _____ cups

Chapter 7
Lesson 4

Halving and Doubling Time and Numbers

NCTM Standards 1, 2, 6, 7, 8, 9, 10
TEKS 2.10B, 2.13A, 2.14

What is missing?

1.

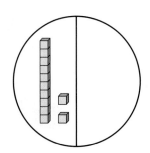

12 is half of __24__.

2.

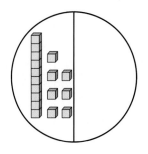

17 is half of _____.

3.

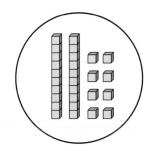

_____ is half of 28.

4.

_____ is half of 56.

	Whole	Half
5.		15
6.	46	
7.		36
8.	58	

NOTE: Your child is learning to find doubles and halves.

How long will a round trip take?

	One Way	Round Trip
9.	9 minutes	18 _____ minutes
10.	15 minutes	_____ minutes
11.	24 minutes	_____ minutes
12.	40 minutes	_____ minutes
13.	55 minutes	_____ minutes

A round trip is double a one-way trip.

 14. How did you find the answer for Problem 13?
Use words, numbers, or pictures to explain.

Challenge

15. What time is the movie half over?
What time will the movie end?

Movie Starts	Half Over	Movie Ends
4:00	:	:

Chapter 7

Lesson 5 # Doubling Length

NCTM Standards 1, 2, 3, 4, 6, 8, 9, 10

🔻 **TEKS 2.9A, 2.13A, 2.14**

What is the distance around each figure?

1.

_____ spaces

2.

_____ spaces

3.

_____ spaces

4.

_____ spaces

5. Draw a figure with 4 sides. What is the distance around the figure?

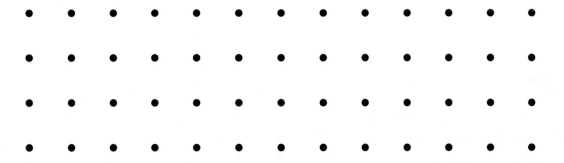

_____ spaces

🏠 **NOTE:** Your child is learning
to double the length of lines.

Draw a new line. Make it twice as long as the blue line. How long is each line?

6.

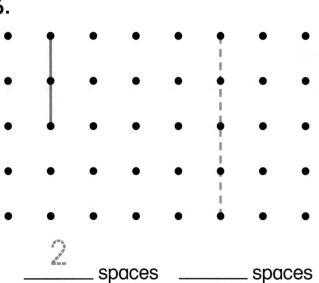

_2_ spaces _____ spaces

7.

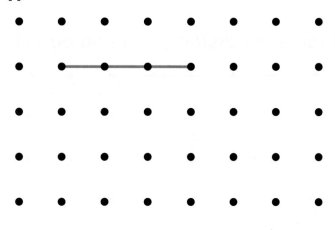

_____ spaces _____ spaces

Draw a new figure. Double the sides of the blue figure.

8.

9.

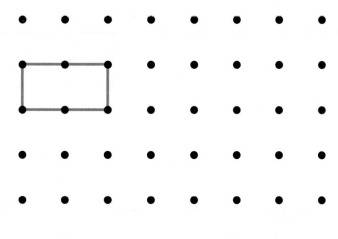

Challenge

10. The distance around a square is 16 spaces. Draw the square. How long is each side?

_____ spaces

Chapter 7
Lesson 6

Thirds and Fourths

NCTM Standards 1, 2, 6, 8, 9, 10
TEKS 2.2A, 2.12A, 2.12B, 2.12C, 2.13A, 2.14

1. Color $\frac{1}{2}$ of the figure.

Use the dots to help you divide the whole into equal parts.

Color $\frac{1}{3}$.

2.

3.

Color $\frac{1}{4}$.

4.

5.

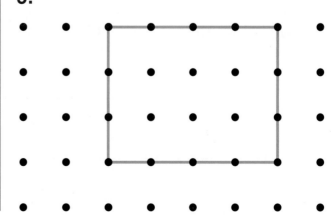

NOTE: Your child is learning to identify and write fractions. Together, fold a napkin to show one third or one fourth.

What part is colored? Write the fraction.

6.

$\frac{1}{3}$

7.

_____

8.

_____

9.

_____

Color $\frac{1}{4}$ in three different ways.

10.

Problem Solving

11. Four friends want to share a whole apple pie. How many pieces do they need to cut the pie into? What fraction of the pie would each person get? Use words, numbers, or pictures to explain.

_____

_____

_____

Chapter 7
Lesson 7

Fair Shares

NCTM Standards 1, 2, 6, 8, 9, 10

TEKS 2.2A, 2.12C, 2.13A, 2.14

Write each fraction.

1.

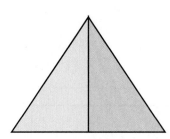

What part is green? $\frac{1}{2}$

What part is blue? _____

2.

What part is green? _____

What part is blue? _____

3.

What part is green? _____

What part is blue? _____

4.

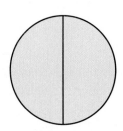

What part is green? _____

NOTE: Your child is learning to identify and write fractions for parts of a whole.

Color to show each fraction.

5.

$\frac{2}{3}$

6.

$\frac{3}{4}$

7.

$\frac{3}{3}$

8.

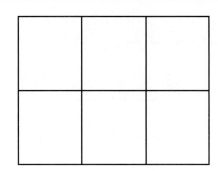

$\frac{4}{6}$

9. Color $\frac{2}{5}$ green. Color $\frac{1}{5}$ blue.

Color the rest red.

What fraction of the figure is red? _____

Problem Solving

10. Erin has 4 stuffed elephants. Three of the elephants are pink. What fraction of the elephants is NOT pink?

Draw a picture to show how you found the answer.

_____

Chapter 7
Lesson 8

Exploring Fractions with Cuisenaire® Rods

NCTM Standards 1, 2, 6, 7, 8, 9, 10

🐾 TEKS 2.2A

The striped rod is one whole. How much is one of the other rods? Write the fraction.

1.

One dark green rod is

_____.

2.

One red rod is

_____.

3.

One purple rod is

_____.

4.

One green rod is

_____.

5.

One white rod is

_____.

🏠 **NOTE:** Your child is learning to represent fractions with Cuisenaire® Rods. Ask your child how many green rods make up $\frac{1}{2}$ of a striped rod.

The top diagram shows a fraction bar model with rows labeled:
- Striped (unlabeled top row)
- D | D
- P | P | P
- G | G | G | G
- R | R | R | R | R | R
- W | W | W | W | W | W | W | W | W | W | W | W

Write <, >, or =. Use the picture to help you.

6. $\frac{1}{3}$ ◯ $>$ $\frac{1}{4}$

P   G

7. $\frac{1}{6}$ ◯ $\frac{1}{4}$

R   G

8. $\frac{1}{6}$ ◯ $\frac{1}{12}$

9. $\frac{1}{6}$ ◯ $\frac{2}{12}$

10. Make your own. Choose a rod in two different colors. Complete the sentence.

$\frac{1}{\square}$ ◯ $\frac{1}{\square}$

Challenge

11. $\frac{1}{4} + \frac{1}{4} + \frac{1}{4}$ ◯ $\frac{1}{3} + \frac{1}{3}$

12. $\frac{1}{3}$ ◯ $\frac{1}{6} + \frac{1}{6}$

Name _____ Date/Time _____

More Fractions

NCTM Standards 1, 2, 6, 7, 8, 9, 10
TEKS 2.2C, 2.13A, 2.14

What part is colored? Write the fraction.

1.

$\frac{5}{9}$

_____

2.

_____

3.

_____

4.

_____

5.

_____

6.

_____

Make your own. Color part of the picture.
Write the fraction you made.

7.

_____

8.

_____

NOTE: Your child is learning
to write fractions and estimate
the relative size of fractions.

Write the fraction for the colored part.
Circle if it is closer to 0, $\frac{1}{2}$, or 1.

9.

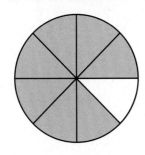

$\frac{7}{8}$

_____

0	$\frac{1}{2}$	(1)

10.

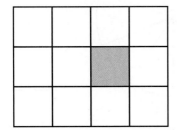

_____

0	$\frac{1}{2}$	1

11.

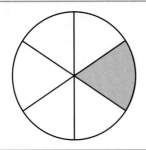

_____

0	$\frac{1}{2}$	1

12.

_____

0.	$\frac{1}{2}$	1

Problem Solving

13. Color the picture to show a fraction close to $\frac{1}{2}$.

   What fraction did you show? _____

   Explain how you know it is close to $\frac{1}{2}$.

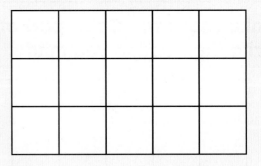

_____

_____

Chapter 7
Lesson 10

Problem Solving Strategy
Guess and Check ✏

NCTM Standards 1, 2, 4, 5, 6, 7, 8, 9, 10
🔽 TEKS 2.12A, 2.12B, 2.12C, 2.12D, 2.13A, 2.14

Understand
Plan
Solve
Check

1. Dan has a cup of marbles. He takes $\frac{1}{2}$ and shares the
   rest fairly between 2 friends. Dan gets 2 more marbles than
   each friend. How many marbles were in the cup? _____ marbles

   How did you find the answer? _____

   _____

   _____

   _____

2. Double a number and it is half of 48.
   What is the number? Explain. _____

   _____

   _____

   _____

   _____

3. There are fewer than 10 pretzels in the bag. If two children
   share, 1 pretzel is left over. If three children share,
   2 pretzels are left over. How many pretzels are in the bag? _____ pretzels

   How did you find the answer? _____

   _____

   _____

 NOTE: Your child is exploring different ways to solve
problems. Sometimes using the strategy, *guess and check,*
is an efficient way to solve a problem.

Problem Solving Test Prep

1. Maria has pennies and dimes in a bag. She picks 3 coins. Which is NOT an amount of money she could have?

  (A) 3¢

  (B) 12¢

  (C) 15¢

  (D) 21¢

2. Sal's team scored 21 points in the second half. They had 50 points at the end of the game. How many points did the team get in the first half?

  (A) 21 points

  (B) 29 points

  (C) 39 points

  (D) 71 points

Show What You Know

3. Nicole rides her bike 1 mile every 10 minutes. She starts riding at 9:00. What time will it be after she rides 4 miles?

Miles					
Time	:	:	:	:	:

Explain your answer.

_____

_____

_____

_____

4. A pizza is cut in 8 pieces. The children eat half a pizza. How many pieces are left?

_____ pieces

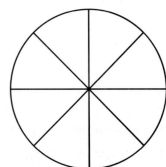

Explain your answer.

_____

_____

_____

_____

Chapter 7 **Review/Assessment**

NCTM Standards 1, 2, 4, 6, 7, 8, 9, 10

Color $\frac{1}{2}$ of each picture. Lesson 1

1.

2.

How much is half of each amount? Lesson 2

3.

Half of 68 is _____.

4.

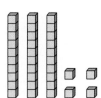

Half of 34 is _____.

How much is double of each amount? Lesson 3

5.

25 doubled is _____.

6.

40 doubled is _____.

7. How long will a round trip take? Lesson 4

One Way	12 minutes	46 minutes	34 minutes
Round Trip	_____ minutes	_____ minutes	_____ minutes

Draw a new figure. Double the sides of the blue figure. Lesson 5

8.

9.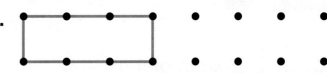

What part is green? Write each fraction. Lessons 6, 7

10.

_____

11.

_____

12.

_____

13. Write the fraction for the colored part.
Circle if it is closer to 0, $\frac{1}{2}$, or 1. Lesson 9

_____

$$0 \qquad \frac{1}{2} \qquad 1$$

Problem Solving Lesson 10

14. Jess has a bag of cherries. She eats half.
Then she gives her brother half of what
is left. She is left with 5 cherries. How many
cherries did she have to start?

_____ cherries

Name _____

Building Addition and Subtraction Fluency
Exploring Related Number Sentences

5 + 7 = 12

How can two number sentences be related?

STEP 1 Observing

Turn over two cards from the pile. Are the number sentences related? Explain.

_____

_____

STEP 2 Observing Some More

Pick another card, and put it with the other two.
Do you see any related sentences now? Explain.

_____

_____

STEP 3 Writing Related Sentences

Pick one of your cards. Write a related number sentence. How are they related?

_____

_____

_____

_____

Investigation

Dear Family,

Today we started Chapter 8 of *Think Math!* In this chapter, I will develop my ability to add and subtract bigger numbers. There are NOTES on the Lesson Activity Book pages to explain what I am learning every day.

Here are some activities for us to do together at home. These activities will help me understand addition and subtraction.

Love,

_____

Family Fun

Race to 100!

Play this game with your child. Your child will play this game in Lesson 6.

■ Prepare two gameboards like this. You also need two number cubes.

I already have...	The sum of the numbers tossed is...	My new total is...
0		

■ Players alternate turns. For each turn, toss the number cubes and add the two numbers. Record the sum in the middle column. Add the first two columns together and record that sum in the third column. Copy the "new total" into the first column of the row for the next turn.

■ The winner is the first player to reach 100 in the "new total" column.

What Makes a Number

Work with your child to find addition problems for a given number.

■ On index cards or slips of paper, write a variety of two-digit numbers, one per card.

■ Place the cards face down in a pile, and have your child choose one card. Together, name two numbers that when added together make the number on the card. Write an addition sentence for your numbers.

51   25 + 26 = 51

■ Repeat, reusing cards if you want. If cards are reused, challenge your child to come up with a different addition sentence for the number.

Name _____   Date/Time _____

Adding with Cuisenaire® Rods

NCTM Standards 1, 2, 4, 6, 7, 8, 9, 10
 TEKS 2.6C, 2.13A, 2.14

What are all the possible 2-car trains?

1.

R

W _____ _____

2.

G

R _____ _____

_____ _____

3.

P

R _____ _____

_____ _____

_____ _____

What are all the possible 3-car trains?

4.

G

W _____ _____ _____

6.

Y

_____ _____ _____

_____ _____ _____

_____ _____ _____

_____ _____ _____

_____ _____ _____

5.

P

R _____ _____ _____

_____ _____ _____

_____ _____ _____

7. How many different rod trains can you make?

Rod Color	2-Car Trains
red	I
green	
purple	

Rod Color	3-Car Trains
green	
purple	
yellow	

 NOTE: Your child is learning to list combinations
and write addition sentences using
Cuisenaire® Rods.

Write an addition sentence for each picture.

8.

   $$W + W = R$$
   _____

9.

   _____

10.

   _____

11.

   _____

12.

   _____

13. What other rod train is as long as a yellow rod?
    Write an addition sentence and draw a picture.

    _____

Challenge

14. Make a 2-car train that is as long as a yellow rod. Then make a new train using two of each car. What color rod matches this longer train? _____

    What addition sentence shows this? _____

Chapter 8
Lesson 2

Exploring Fact Families

NCTM Standards 1, 2, 6, 7, 8, 9, 10
TEKS 2.5C, 2.14

What is the fact family for each picture?

1.

R + P

_____ + _____ = _____

_____ + _____ = _____

_____ − _____ = _____

_____ − _____ = _____

2.

_____ + _____ = _____

_____ + _____ = _____

_____ − _____ = _____

_____ − _____ = _____

3.

_____ + _____ = _____

_____ + _____ = _____

_____ − _____ = _____

_____ − _____ = _____

NOTE: Your child is learning to write a fact family of addition and subtraction sentences for Cuisenaire® Rod trains.

4. What is the fact family for these rod trains?

_____ + _____ = _____

_____ + _____ = _____

_____ − _____ = _____

_____ − _____ = _____

---

 5. Write the other sentences for this fact family.
Tell how you completed the sentences.

R + G = Y

_____ − _____ = _____

_____ + _____ = _____   _____ − _____ = _____

_____

_____

_____

_____

Challenge

6. Use (+ ☺) = △ to complete each sentence.

☺ + (= _____   △ − ☺ = _____

△ − (= _____

Chapter 8
Lesson 3

Connecting Addition and Subtraction

NCTM Standards 1, 2, 6, 7, 8, 9, 10
TEKS 2.1A, 2.3A, 2.3B, 2.5C

What is missing in each fact family?
Draw the symbols. Write the numbers.

1.

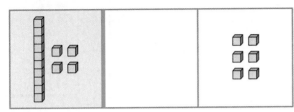

14 ____ – ____ = 6

____ – ____ = ____

____ + ____ = ____

____ + ____ = ____

2.

 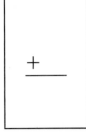

15
+
20

+

–

–

3.

____ = ____ + ____

____ = ____ + ____

____ = ____ – ____

____ = ____ – ____

4.

 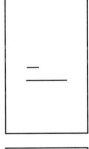

–

–

+

+

NOTE: Your child is learning to write fact families that relate addition and subtraction problems.

What is missing in each fact family?
Draw the symbols. Write the numbers.

5.

_____ − _____ = _____

_____ − _____ = _____

_____ + _____ = _____

_____ + _____ = _____

6.

+ _____

+ _____

− _____

− _____

✏️ **7.** Pick a number sentence from Problem 6.
Write a story to match it.

_____

_____

_____

_____

Challenge

8. Write the fact family.

_____ = _____ + _____   _____ = _____ − _____

_____ = _____ + _____   _____ = _____ − _____

Name _____ Date/Time _____

Adding and Subtracting Using 5 and 10

NCTM Standards 1, 6, 7, 8, 9, 10

TEKS 2.3A, 2.3B, 2.3C, 2.3D, 2.5C, 2.14

How much is each collection worth?

1.

_____¢

2.

_____¢

What is missing? Complete the expressions for each number.

3. | 8 |

5 + __3__

4 + __4__

10 − __2__

4. | 14 |

5 + _____

7 + _____

10 + _____

5. | 15 |

5 + _____

5 + 5 + _____

20 − _____

6. | 21 |

20 + _____

1 + 10 + _____

10 + _____

7. | 25 |

5 + _____

10 + 10 + _____

15 + _____

8. | 36 |

26 + _____

30 + _____

16 + _____

NOTE: Your child is learning a strategy to help add and subtract by breaking numbers apart into fives and tens.

Add or subtract.

9. $8 + 21 =$ _____	**10.** $14 + 9 =$ _____	**11.** $15 + 25 =$ _____
12. $14 - 6 =$ _____	**13.** $21 - 8 =$ _____	**14.** $36 - 25 =$ _____

15.   27¢
    + 5¢
      ¢
	16. 40¢
	   − 15¢
	      ¢
		17. 25¢
		   − 9¢
		      ¢

18. Use words, numbers, or pictures to explain how you solved Problem 17.

Problem Solving

19. Barbara has 35¢. She buys a whistle for 10¢ and a sticker for 8¢. How much money does she have left? Explain. _____¢

_____

_____

_____

_____

Adding and Subtracting Numbers Near 10

NCTM Standards 1, 2, 6, 7, 8, 9, 10

TEKS 2.3B, 2.6B, 2.6C, 2.13A, 2.14

What is missing in each table? Use the rule.

1.

a	11	12	21	22	30		32	43	
$a + 8$	19	20	29			39			52

2.

n	29	28	50	51		61		91	
$n - 9$	20	19			43		62		83

3.

10	20		40	90		92	73		c
2		22			83			66	$c - 8$

4.

28	27	78	77	58	57	37		87	m
37		87					26		$m + 9$

NOTE: Your child is learning to solve addition and subtraction problems by using 10 and then adjusting the results.

What is missing? Complete each sentence.

5. 37 + 8

37 + 10 = 47

47 − 2 = ___ 45

6. 28 + 9 = _____

7. 12 + 49 = _____

8. 64 + 11 = _____

Find convenient numbers to help you add and subtract.

9. 42 − 8

42 − 10 = □

□ + 2 = _____

10. 33 − 9 = _____

11. 51 − 12 = _____

12. 25 − 11 = _____

 13. How would you solve 54 − 9? Use words, numbers, or pictures to explain.

Challenge

14. Complete the table in two different ways. Write the rules.

a	6	15	23	47	58
	12				

a	6	15	23	47	58
	12				

---

168 one hundred sixty-eight   **CLXVIII**   14 dozen

Chapter 8
Lesson 6

Place Value and Cross Number Puzzles

NCTM Standards 1, 2, 6, 8, 9, 10

🔻 TEKS 2.1B, 2.3B

What is missing? Write the numbers or the symbols.

1.

2.

10		17
20	5	25

3.

4.

	4	
10		16
20		

5.

20		21
		45
	3	13
70	9	

🏠 **NOTE:** Your child is learning to add and subtract numbers using Cross Number Puzzles. Ask your child to explain how to solve Problem 5.

What is missing? Write the numbers.

6.

	0	30
0		9
	8	18
40		

8.

10	5	
	4	
20		28
80		

7.

	7	
0	6	
40		42
80	15	

9.

31		1
	10	8
29		
		18

Challenge

10. Fill in the Cross Number Puzzle to help solve this problem.

$$\begin{array}{r} 86 \\ +\ 47 \\ \hline \end{array}$$

Name _____ Date/Time _____

Breaking Numbers Apart

NCTM Standards 1, 2, 6, 7, 8, 9, 10

 TEKS 2.3A, 2.3B

Break the numbers in different ways so they are easier to add. Write the missing numbers.

> 5 + 3, 1 + 7, and 10 − 2 are all different ways to show 8.

1.

| 8 + 7 | ⟹ | 5 + 3 + 5 + 2 ——— + 5 | ⟹ | 1 + 7 + 4 + 3 ——— + | ⟹ | 10 − 2 + 10 − 3 ——— − |

2.

| 15 + 18 | ⟹ | 10 + 5 + 10 + 8 ——— + | ⟹ | + 10 + 5 + ——— + | ⟹ | 20 − + − 2 ——— − 7 |

3.

| 9 + 34 | ⟹ | 4 + + 9 + ——— 13 + | ⟹ | 9 + + 31 + ——— + | ⟹ | 10 − + − 6 ——— − |

4.

| 26 + 26 | ⟹ | + 1 + + 1 ——— + 2 | ⟹ | 20 + + 20 + ——— + | ⟹ | 20 − + − 6 ——— − |

 NOTE: Your child is learning strategies that make it easier to solve addition and subtraction problems.

Break the numbers apart in different ways.
Then complete the puzzle. Circle the way
that matches the puzzle.

5.

| 14 |
| + 28 |

$\Rightarrow$

| + 4 |
| + 20 + |
| ___ |
| + |

$\Rightarrow$

| 7 + |
| + 14 + |
| ___ |
| + |

$\Rightarrow$

| 20 − |
| +   − 2 |
| ___ |
| − 8 |

10		14
	8	28
	12	

6.

| 76 |
| + 28 |

$\Rightarrow$

| 75 + |
| + 25 + |
| ___ |
| + |

$\Rightarrow$

| + 6 |
| + 20 + |
| ___ |
| + |

$\Rightarrow$

| 80 − |
| +   − 2 |
| ___ |
| − |

70		
	8	
90		

7.

| 37 |
| + 19 |

$\Rightarrow$

| 20 + |
| + 16 + |
| ___ |
| + |

$\Rightarrow$

| 40 − |
| +   − 1 |
| ___ |
| − |

$\Rightarrow$

| + 7 |
| + 10 + |
| ___ |
| + |

30	7	
10		19

Challenge

8. Break the numbers apart in different ways.

| 149 |
| + 46 |

$\Rightarrow$

| +   + 100 |
| + 6 +   +   0 |
| ___ |
| +   + |

$\Rightarrow$

| 150 − |
| +   − 4 |
| ___ |
| − |

Name _____   Date/Time _____

Using Cross Number Puzzles to Subtract

NCTM Standards 1, 2, 6, 7, 8, 9, 10

TEKS 2.3B, 2.3C, 2.14

What is missing?

1.

40		52
20	4	24

2.

71	60	11
46		6

3.

76		
	80	13

4.

	9	59
70		83

Show how to solve each problem.

5.
$$\begin{array}{r} 82 \\ -\ 69 \\ \hline \end{array}$$

6.
$$\begin{array}{r} 38 \\ +\ 47 \\ \hline \end{array}$$

NOTE: Your child is learning to solve addition and subtraction problems using different strategies.

What is missing?

7.
$$
\begin{array}{r}
26 \\
+\ 46 \\
\hline
\end{array}
$$

8.
$$
\begin{array}{r}
62 \\
-\ 30 \\
\hline
\end{array}
$$

Think about the numbers in a problem. What method would work the best?

9. $34 - 19 =$ _____

10. $51 + 25 =$ _____

11. $98 + 35 =$ _____

12. $196 - 97 =$ _____

13. How would you solve $53 - 29$? Use words, numbers, or pictures to explain.

Problem Solving

14. Lenny read 49 pages of his book on Monday. He read 23 pages on Tuesday. How many pages did he read on both days?

_____ pages

15. Pia's book is 302 pages long. She has read 150 pages. How many more pages does Pia have left to read?

_____ pages

Chapter 8
Lesson 9

Comparing Mathematical Expressions

NCTM Standards 1, 2, 6, 7, 8, 9, 10
TEKS 2.1C, 2.5B, 2.14

Make each sentence true. Write >, <, or =.

1. $64 + 36$ $\bigcirc$ $64 + 37$
   Answer: $<$

7. $36 - 26$ $\bigcirc$ $86 - 76$

2. $54 - 5$ $\bigcirc$ $54 - 10$

8. $57 - 37$ $\bigcirc$ $46 - 36$

3. $84 + 37$ $\bigcirc$ $37 + 84$

9. $24 + 36$ $\bigcirc$ $25 + 35$

4. $87 - 48$ $\bigcirc$ $87 - 50$

10. $64 - 39$ $\bigcirc$ $65 - 39$

5. $72 + 43$ $\bigcirc$ $42 + 72$

11. $64 - 39$ $\bigcirc$ $65 - 40$

6. $39 + 12$ $\bigcirc$ $13 + 39$

12. $39 + 13$ $\bigcirc$ $12 + 40$

NOTE: Your child is learning to compare addition and subtraction expressions using <, >, and =.

What is missing?

13. $28 + 32 = \underline{} + 28$

17. $30 + 28 = 20 + \underline{}$

14. $43 - \underline{} = 43 - 19$

18. $47 - 18 = 57 - \underline{}$

15. $15 + 17 = 16 + \underline{}$

19. $24 - 12 = \underline{} - 13$

16. $\underline{} + 63 = 12 + 64$

20. $62 + 36 = \underline{} + 60$

 21. How did you solve Problem 20? Use words, numbers, or pictures to explain.

Creating and Solving Story Problems

NCTM Standards 1, 2, 6, 7, 8, 9, 10
🔸 TEKS 2.3C, 2.13A, 2.14

> Jamie sold 63 red balloons, 48 blue
> balloons, and 24 green balloons.

1. Circle a question that can be answered from the story.

- How many balloons did Jamie sell?

- How much money did Jamie make?

- How many more red balloons than blue balloons did Jamie sell?

- How much string did Jamie use?

2. Write a number sentence to match the question you circled.

3. Show how you would solve the problem.

NOTE: Your child is learning to work with addition and
subtraction story problems. Ask him or her to make up a
story problem for you to solve.

Which number sentence matches each story?

4.

Ms. Lee buys 73 notebooks. Mr. Hall buys the same number of notebooks.

$73 + 27 =$ _____

5.

Andrew collected 73 shells. He used 27 of the shells to make a frame. How many shells are not in the frame?

$73 =$ _____

6.

Tamika had 73 crayons in her box. She found 27 more. Now how many crayons does she have?

$73 - 27 =$ _____

Challenge

7. Solve Problem 6. Show how you solved it.

Chapter 8
Lesson 11

Strategies for Multiple-Choice Questions

NCTM Standards 1, 6, 8, 9, 10
TEKS 2.3C

Fill in the bubble for each correct answer.

It helps to cross out answers that do not make sense for a problem.

1.   13
       + 17

Color in the correct answer.

Ⓐ 4   Ⓒ 30
Ⓑ 20   Ⓓ 45

---

2.   3
           + 7

Ⓐ 4
Ⓑ 8
Ⓒ 9
Ⓓ 10

3. 30 − 10 = _____

Ⓐ 20   Ⓒ 40
Ⓑ 31   Ⓓ 300

---

4.   43
           − 28

Ⓐ 71
Ⓑ 61
Ⓒ 25
Ⓓ 15

5. 25 + 17 = _____

Ⓐ 8   Ⓒ 42
Ⓑ 32   Ⓓ 312

 NOTE: Your child is learning to answer multiple-choice questions. Ask him or her how to eliminate answer choices to help solve the problems.

Fill in the bubble for each correct answer.

6.
$$\begin{array}{r} 49 \\ + 97 \\ \hline \end{array}$$

(A) 1,316

(B) 154

(C) 146

(D) 50

7.
$$\begin{array}{r} 56 \\ - 19 \\ \hline \end{array}$$

(A) 37

(B) 40

(C) 43

(D) 75

8. $175 + 20 =$ _____

(A) 195

(C) 165

(B) 177

(D) 155

9. $315 + 9 =$ _____

(A) 325

(C) 306

(B) 324

(D) 225

BIG TOY SALE!
trains 12¢   cars 35¢   boats 25¢

10. Emma bought a toy car. She paid the clerk 50¢. How much change did she get?

(A) 15¢

(C) 38¢

(B) 25¢

(D) 40¢

11. Dane bought a train. Ciara bought a boat. How much more did Ciara spend than Dane?

(A) 9¢

(C) 23¢

(B) 13¢

(D) 37¢

Challenge

12. Brad bought 3 toy cars and 1 train. Claudia bought 6 boats. How much more did Claudia spend than Brad?

(A) 47¢

(C) 43¢

(B) 45¢

(D) 33¢

Problem Solving Strategy
Solve a Simpler Problem

NCTM Standards 1, 2, 6, 7, 8, 9, 10
TEKS 2.12B, 2.12C

Understand
Plan
Solve
Check

1. Martha had 57 stamps in her collection. She
   bought a package of 15 stamps. Then how many
   stamps did she have in the collection?

   _____ stamps

   How did you find the answer?

   _____

   _____

2. There were 184 magazines in the store. Another
   55 magazines were delivered. Then the store sold
   50 magazines. How many magazines were in
   the store then?

   _____ magazines

   How did you find the answer?

   _____

   _____

3. Howie spent 16 minutes on his math
   homework and 25 minutes on his reading
   homework. How much time did Howie
   spend doing his homework?

   _____ minutes

   How did you find the answer?

   _____

   _____

 NOTE: Your child is exploring different ways to solve
problems. Sometimes solving a simpler problem is an
efficient way to find the answer.

Problem Solving Test Prep

1. Kyle rides his bike 1 mile every 10 minutes. He starts riding at 9:00. How far will he ride by 9:30?

   Ⓐ 1 mile

   Ⓑ 2 miles

   Ⓒ 3 miles

   Ⓓ 4 miles

2. Half of a number is between 20 and 25. What number could it be?

   Ⓐ 11

   Ⓑ 24

   Ⓒ 46

   Ⓓ 55

Show What You Know

3. Tina makes a bead necklace. She strings on 1 white bead, 2 blue beads, 1 white bead, 2 blue beads. If she continues this pattern, what color will the eighth bead be?

   _____

   Explain how you found the answer.

   _____

   _____

   _____

4. The Eagle basketball team scored 16 points in the second half of the game. They had 38 points at the end of the game. How many points did the team score in the first half?

   _____ points

   Explain how you know.

   _____

   _____

   _____

Chapter **8** Review/Assessment
NCTM Standards 1, 2, 6, 8, 9, 10

What is missing in each fact family? Draw the symbols or write the numbers. Lessons 1–3

1.

_____ + _____ = _____

_____ + _____ = _____

_____ − _____ = _____

_____ − _____ = _____

2.
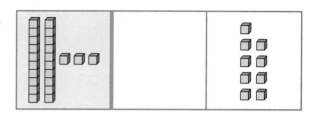

_____ − _____ = _____

_____ − _____ = _____

_____ + _____ = _____

_____ + _____ = _____

What is missing? Complete the expressions for each number. Lesson 4

3. | 12 |

10 + _____

6 + _____

5 + _____

4. | 34 |

30 + _____

24 + _____

14 + _____

5. What is missing in the table? Use the rule. Lesson 5

m	13	14	50		71	72	38		49
$m + 11$	24		61	60				50	

What is missing? Complete each puzzle. Lessons 6–8

6.

	2	
40		45
70		

7.

60	11	71
	3	53

---

8. Break the numbers apart. Circle the box that matches the puzzle. Lesson 7

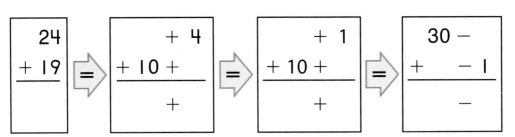

$$\begin{array}{r} 24 \\ + 19 \end{array}$$

$$\begin{array}{r} + 4 \\ + 10 + \\ \hline \\ + \end{array}$$

$$\begin{array}{r} + 1 \\ + 10 + \\ \hline \\ + \end{array}$$

$$\begin{array}{r} 30 - \\ + \quad - 1 \\ \hline \\ - \end{array}$$

20		
	9	19
	13	

Make each sentence true. Write >, <, or =. Lesson 9

9. $21 + 38 \bigcirc 20 + 38$

10. $45 - 13 \bigcirc 45 - 23$

Fill in the bubble for each correct answer. Lesson 11

11.
$$\begin{array}{r} 39 \\ + 25 \\ \hline \end{array}$$

Ⓐ 514
Ⓑ 64
Ⓒ 54
Ⓓ 14

12.
$$\begin{array}{r} 87 \\ - 38 \\ \hline \end{array}$$

Ⓐ 125
Ⓑ 59
Ⓒ 49
Ⓓ 45

Problem Solving Lesson 12

13. Ms. Chen sold 23 apple muffins at the bake sale. She sold 41 corn muffins. How many corn and apple muffins did she sell?

_____ muffins

Chapter 9

Two-Dimensional Figures and Spatial Sense
The Shape of Signs

Create and sort signs.

STEP 1 **Making a Sign**

Draw a sign on a sheet of paper.

How many sides does your sign have? _____

How many corners does your sign have? _____

STEP 2 **Sorting the Signs**

Sort the signs in any way. Use words to tell how you sorted.

_____

_____

STEP 3 **Sorting Another Way**

Tell another way you could sort the signs.

_____

_____

_____

_____

_____

_____

Investigation

School-Home Connection

Dear Family,

   Today we started Chapter 9 of *Think Math!* In this chapter, I will explore two-dimensional figures, symmetry, area, and how to record paths. There are NOTES on the Lesson Activity Book pages to explain what I am learning every day.

   Here are some activities for us to do together at home. These activities will help me understand two-dimensional figures and symmetry.

Love,

_____

Family Fun

I Spy

Work with your child to play a game about two-dimensional figures, called *I Spy*.

- Review the names of these figures.

   square   **rectangle**   **triangle**

- To play the game, you secretly choose an object in the room that is (or contains) one of these figures. Then you say, "I spy a square (rectangle or triangle)."

- Your child asks *yes/no* questions about the object until he or she guesses the correct one. Some good questions to ask are: "Is the square in a place that we can easily see?" or "Does the object have more than one square?"

- Take turns selecting an object and asking the questions.

Lines of Symmetry

Work with your child to make a figure that has a line of symmetry.

- You will need paper and scissors. Help your child fold the sheet of paper in half. Make sure the two halves match.

- Help your child draw a design that starts and ends at the fold. Together, guess what the shape will look like when it is cut out and the paper is opened.

- Hold the folded side of the paper and cut out the design. You should not cut along the fold.

- Invite your child to unfold the paper and draw a line down the fold. You made a line of symmetry!

Sorting Polygons by Attributes

NCTM Standards 2, 3, 6, 7, 8, 9, 10
TEKS 2.7A, 2.7B, 2.13A

Count the polygons in the picture.
Complete each sentence.

1. There are ___18___ polygons in the picture.

_____

2. There is _____ polygon with 3 sides in the picture.

_____

3. There are _____ polygons with 4 sides in the picture.

_____

4. There are _____ polygons with more than 4 sides.

_____

NOTE: Your child is learning about
polygons. A polygon is a closed figure
with all straight sides.

5. Draw a design using only polygons. Use a straightedge.
Start and end each line at a dot.

Problem Solving

6. I have 4 straight sides. I am closed
and have no curves. All of my sides
are the same length. What figure am I?

_____

Chapter 9
Lesson 2

Congruent and Similar Figures

NCTM Standards 3, 4, 6, 7, 8, 9, 10
TEKS 2.7B, 2.12A, 2.13A

1. Draw the same polygon in different positions.
   Use 4 squares for each polygon. Draw as many
   congruent polygons as you can.

The figures are all congruent.

---

2. Draw a figure congruent to this one.

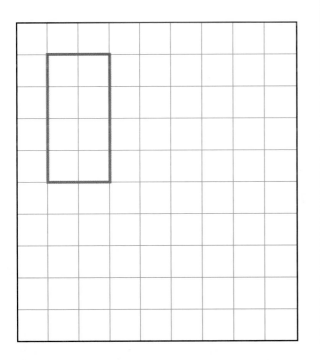

3. Draw 2 congruent figures of your own.

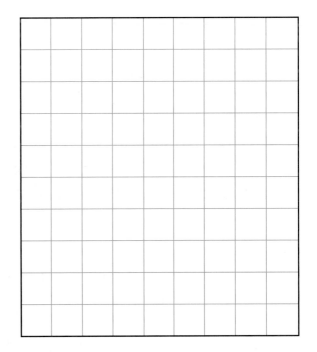

NOTE: Your child is learning about congruent and similar
figures. Congruent figures are the same size and shape.
Similar figures may be different sizes.

Circle the pairs of figures that are similar.
Put an X on the pairs that are *not* similar.

4.

5.

6.

7.

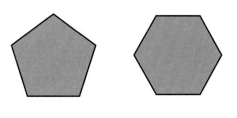

8. Draw 2 figures similar to this one.

Problem Solving

9. Robin says that she can make a square with
5 congruent squares. Is she correct?
Use words, numbers, or pictures to explain.

Name _____   Date/Time _____

Building with Triangles

NCTM Standards 2, 3, 4, 6, 7, 8, 9, 10

TEKS 2.7A, 2.7C, 2.13A

Draw lines in each figure to make 4 small congruent triangles.

Congruent triangles are the same size and shape.

1.

2.

 NOTE: Your child is learning to make congruent and similar triangles. Similar triangles have the same shape, but do not have to be the same size.

3. Draw a similar triangle on the grid.

✏️ **4.** What is the same about these 2 triangles?
What is different?

_____

_____

_____

_____

Challenge

5. Draw a similar figure on the grid.
Use 4 of the smaller figures.

Chapter 9
Lesson 4 # Looking at Reflections

NCTM Standards 3, 6, 7, 8, 9, 10

TEKS 2.13A

The thick green line shows where to place the mirror.

Draw the reflection of each figure. Use a mirror to help you.

1.

2.

3.

4.

5.

6.

7.

8.

9.

10.

 NOTE: Your child is learning to draw reflections of letters and figures. Together, hold a page from a book in front of a mirror and try to read the words.

Draw the reflections. Work across and down within each grid.

11.

12.

13.

14.
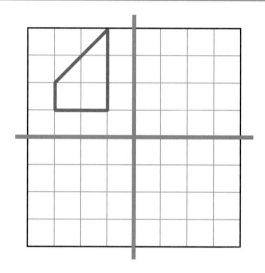

Challenge

15. How many different places can you put a mirror on the square and see the whole square? Draw a line where you put the mirror each time.

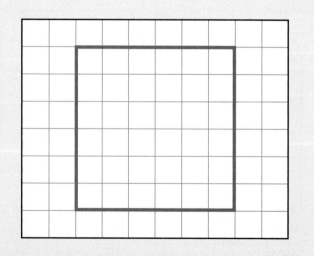

_____ places

Chapter 9
Lesson 5 # Lines of Symmetry

NCTM Standards 2, 3, 6, 7, 8, 9, 10
TEKS 2.13A

Look at each picture. Can you divide it into 2 matching parts? If so, draw all the lines of symmetry. If not, write *no*.

1.

2.

3.

4.

5.

6.

7.

8.

9.

NOTE: Your child is learning about symmetry.
Together, look for objects at home that have
two identical matching parts when folded on a line.

Draw the other half of each picture.

10.

11.

12.

13.

Challenge
Draw a picture for each line of symmetry.

14.

15.

Chapter 9
Lesson 6

Cutting Polygons Apart

NCTM Standards 2, 3, 6, 7, 8, 9, 10
TEKS 2.7B, 2.7C, 2.12A, 2.13A

What new figures do you get if you cut along the lines?

1.

triangles
_____

2.

_____

3.

_____

4.

_____

5. Cut the rectangle into 2 congruent pieces.
Show 2 different ways.

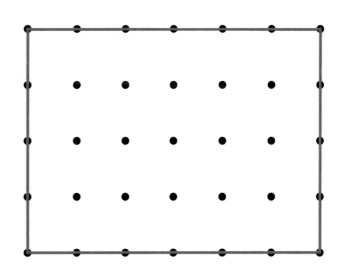

NOTE: Your child is learning to cut polygons to make other polygons. Together, try to cut a sheet of paper into 2 triangles.

Draw a line or lines to show the new figures.

6. Make 2 rectangles.

7. Make 2 triangles.

8. Make 4 triangles.

9. Make 2 triangles and 2 rectangles.

10. Mary cut the hexagon from vertex to vertex to make 2 pieces. How many sides would each piece have?

_____ sides

_____ sides

Problem Solving

11. Raffi drew 4 lines that cut her rectangle into only triangles. How many triangles did she make? Draw 2 different ways to do it.

_____ triangles

_____ triangles

Measuring Area

NCTM Standards 1, 2, 3, 4, 6, 7, 8, 9, 10
TEKS 2.9B, 2.13A

What is the area? Each ☐ is 1 square unit.

1.

_____8_____ square units

2.

_____ square units

3.

_____ square units

4.

_____ square units

5.

_____ square units

6.

_____ square units

NOTE: Your child is learning to find the area of
a polygon by counting the square units inside
the figure.

Draw each figure.

7. a polygon of 5 square units

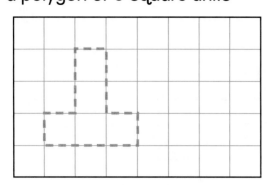

8. a polygon of 6 square units

9. a polygon of 3 square units

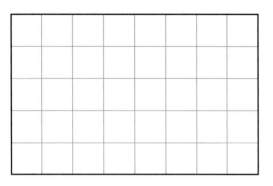

10. a polygon of 8 square units

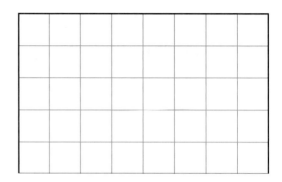

11. a polygon of 9 square units with 8 sides

12. a polygon of 5 square units with 8 sides

Challenge

13. Draw a polygon with an area of $2\frac{1}{2}$ square units. Explain how to count the area of your polygon.

_____

_____

Chapter 9
Lesson 8
Recording Paths

NCTM Standards 2, 4, 6, 7, 8, 9, 10
TEKS 2.12A, 2.12C, 2.13A

What are the shortest paths from dot to dot?
Find all possible ways.

1. from **A** to **B**

A •———• B

Go 1 block East.

_____

2. from **C** to **D**

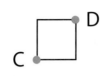

___ ___ ___

___ ___ ___

3. from **E** to **F**

___ ___ ___

___ ___ ___

___ ___ ___

4. from **G** to **H**

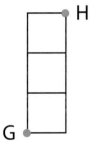

___ ___ ___

___ ___ ___ ___

___ ___ ___ ___

___ ___ ___

5. from **I** to **J**

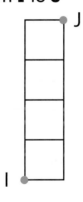

___ ___ ___ ___

___ ___ ___ ___

___ ___ ___ ___

___ ___ ___ ___

6. Complete the table.

Path	Number of Ways
A to **B**	
C to **D**	
E to **F**	
G to **H**	
I to **J**	

NOTE: Your child is learning to describe paths on simple maps using notation such as N, S, E, and W to represent the directions North, South, East, and West.

What is a shortest path from place to place?

N stands for North.

7. from home to the store ___ ___ ___ ___ ___

8. from the store to school ___ ___ ___ ___ ___

9. from the park to home ___ ___ ___ ___ ___

10. Choose your own starting and ending places.
Tell what path you will take.

from _____ to _____ ___ ___ ___ ___ ___

Problem Solving

11. Tina goes to the office and the store.
She starts and ends at home. If Tina takes
a shortest path, how many blocks does
she walk in all? Explain.   _____ blocks

_____

_____

_____

_____

Chapter 9
Lesson 9

Directions from Here to There

NCTM Standards 2, 3, 6, 7, 8, 9, 10
TEKS 2.12A, 2.12C, 2.13A

What are the shortest paths from place to place?
Write the shorthand that describes them all.

6N 2W

1. from Anna's house to Miya's house _____

2. from Chandra's house to Andre's house _____

3. from Andre's house to Oscar's house _____

4. from Anna's house to Andre's house _____

5. Draw a dot anywhere on the map to show
   a new house. What is the shortest path from
   Oscar's house to the new house? _____

NOTE: Your child is learning to describe
relative position on a map using the directions
north, south, east, and west.

Use the clues to mark each place on the map.

home

6. **the store**

The shortest paths from home to the store are 4W 2S.

7. **the school**

The shortest paths from home to the school are 5E 1N.

8. **the park**

The shortest paths from the school to the park are 5S 8W.

9. **the library**

The shortest paths from home to the library are 2S 3E.

 10. Compare the shortest paths from home to the store and then from the store to home. What can you say about the lengths of the 2 paths?

_____

_____

_____

_____

Challenge

11. The bank is 4 blocks away from the store. It is 4 blocks away from home. It is 5 blocks away from the park. Mark the bank on the map.

Problem Solving Strategy
Draw a Picture ✏️
NCTM Standards 1, 2, 3, 6, 7, 8, 9, 10
👉 TEKS 2.12A, 2.12B, 2.12C, 2.13A

Understand
Plan
Solve
Check

1. Eve has 3 congruent square tiles. She wants to build a polygon with just 1 line of symmetry. What polygon can she make?

   _____

   _____

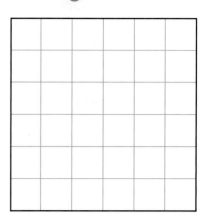

2. Bob wants to cut a rectangle into 2 congruent pieces. What figures can he make?

3. Lisa has 6 game cards. Half of the cards have hearts on them. The rest of the cards have flowers on them. How many of the cards have flowers on them?

   _____ cards

NOTE: Your child is exploring different ways to solve problems. Sometimes drawing a picture is an efficient way to solve a problem.

Problem Solving Test Prep

1. Marie gives $\frac{1}{4}$ of her cookies to Matt. She keeps the 6 cookies that are left. How many cookies did Marie have to start?

   (A) 4 cookies

   (B) 8 cookies

   (C) 12 cookies

   (D) 16 cookies

2. The chef cooks 5 pancakes every 2 minutes. How many pancakes would he make after 10 minutes?

Minutes	2	4	6	8	10
Pancakes	5				

   (A) 9 pancakes

   (B) 10 pancakes

   (C) 20 pancakes

   (D) 25 pancakes

✎ Show What You Know

3. Fran has 28¢. What is the smallest number of coins she could have?

   _____ coins

   Explain how you know that this is the smallest number.

   _____

   _____

   _____

   _____

4. Mike can fit about 25 beans in his hand. He puts the beans in a cup. About how many beans might fill the cup?

   about _____ beans
   Explain how you made your estimate.

   _____

   _____

   _____

Chapter 9 **Review/Assessment**

NCTM Standards 2, 3, 4, 6, 7, 8, 9, 10

1. Circle the figures that are polygons. Lesson 1

2. Are the figures similar? Write **yes** or **no.** Lesson 3

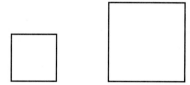

_____

3. Circle the figure congruent to the blue triangle. Lesson 2

  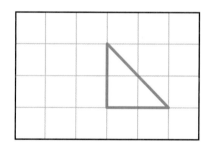

Draw the reflection of each figure. Use a mirror to help you. Lesson 4

4.

5.

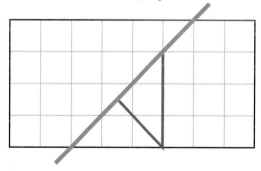

Can you divide each picture into 2 matching parts? If so, draw as many lines of symmetry as you can. If not, write _no._ Lesson 5

6.

7.

What new figures do you get if you cut along the lines? Lesson 6

8.

_____

9.

_____

What is the area? Each ☐ **is I square unit.** Lesson 7

10.

_____ square units

11.
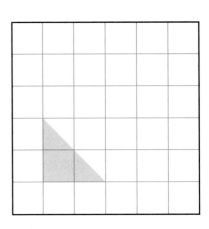

_____ square units

What are the shortest paths from L to M? Write the shorthand. Lessons 8, 9

12.

_____

13.
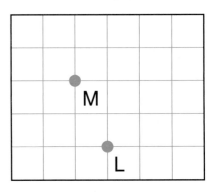

_____

Problem Solving Lesson 10

14. Kelly wants to cut a square into 3 congruent pieces. What figures can she make? Draw a picture.

Chapter 10 Adding and Subtracting Larger Numbers

Double Your Number ✏️

TEKS 2.5B

You need
• index cards

STEP 1 Reading and Writing Numbers

Pick a card from the pile. Read the numbers on both sides of the card aloud. Write them here.

_____   _____

STEP 2 Comparing

How are both numbers alike? _____

_____

How are they different? _____

_____

_____

_____

STEP 3 Doubling the Numbers

Try to double each number. Show what you do.

_____

_____

_____

_____

Investigation

Dear Family,

Today we started Chapter 10 of *Think Math!* In this chapter, I will develop my ability to add and subtract bigger numbers. There are NOTES on the Lesson Activity Book pages to explain what I am learning every day.

Here are some activities for us to do together at home. These activities will help me understand addition and subtraction.

Love,

_____

Family Fun

Counting Coins

Work with your child to play this game. Your child will play a similar game in Lesson 2.

- You will need a recording sheet, like the one shown below, and a cup with at least 5 pennies, 5 nickels, 5 dimes, and 5 quarters.

Q	D	N	P	Total Value	Fewest Coins?
1	2	1	1	51¢	no

- You and your child take turns picking any 5 coins from the cup and recording in the table.

- Is your amount shown with the fewest coins? If not, then trade to show the fewest coins. Return the coins to the cup each time. Shake the cup before each turn.

- Play until you and your child each take 5 turns.

Shopping for Bargains

Work with your child to identify money amounts in dollar and cents notation.

- Look through store flyers listing prices for items in dollar and cents notation.

- Have children look for patterns in the prices. For example, many of the prices have 99 cents, such as $5.99 or $499.99.

TV CITY

$399.99

- Together, make up stories about going to the store to buy one of the items.

Making Sums of 100

NCTM Standards 1, 2, 6, 7, 8, 9, 10
TEKS 2.3B, 2.13A, 2.14

What number is missing?

1.

$$25$$
$$+ \boxed{75}$$
$$\overline{100}$$

2.

$$\boxed{}$$
$$+ \quad 55$$
$$\overline{100}$$

3.

$$23$$
$$+ \boxed{}$$
$$\overline{100}$$

4.

$$81$$
$$+ \boxed{}$$
$$\overline{100}$$

5. $92 + \boxed{} = 100$

6. $14 + \boxed{} = 100$

7. $\boxed{} + 48 = 100$

8. $\boxed{} + 66 = 100$

9. Make your own.

$\boxed{} + \boxed{} = 100$

NOTE: Your child is learning to find number pairs
with a sum of 100. Ask your child to find the number
that when added to 32 makes a sum of 100.

10. Which pairs make 100? Circle them as fast as you can.

Sums of 100 Search

36 / 64	68 / 43	60 / 50	51 / 49	30 / 75	80 / 20	12 / 93	51 / 64	27 / 73
60 / 40	65 / 10	28 / 72	85 / 15	60 / 30	74 / 54	29 / 80	32 / 68	50 / 50
76 / 34	90 / 10	45 / 55	37 / 75	17 / 83	70 / 20	62 / 38	93 / 16	21 / 79

11. Pick one of the numbers pairs you just circled. How do you know that it has a sum of 100?

_____

_____

_____

_____

Challenge

12. Complete the addition sentence. Then write two subtraction sentences related to it.

22 + ☐ = 100

☐ − ☐ = ☐

☐ − ☐ = ☐

Name _____   Date/Time _____

Adding with Coins

NCTM Standards 1, 2, 6, 7, 8, 9, 10
TEKS 2.3D, 2.13A, 2.14

What is the value for each collection of coins?

1.

35
_____ ¢

2.

_____ ¢

3.

_____ ¢

4.

_____ ¢

Show each amount of money with the fewest coins.

5. 35¢

6. 62¢

7. 59¢

8. 40¢

🏠 **NOTE:** Your child is learning to find
the value of a collection of coins and to use
the fewest coins for that value.

What is missing? Complete the table.

Number of Coins	Q	D	N	P	Total Value	Is it the fewest coins? If not, draw a picture.
9. 5	1	0	3	1	41¢	Ⓠ Ⓝ Ⓓ Ⓟ
10. 8	0	5		3		
11.					35¢	yes
12. 9			3	5	45¢	
13.	2			5	60¢	
14.	0	2	1	4		

Problem Solving

15. Kate has some coins worth 51¢. She could NOT have the fewest coins for that amount. What coins might Kate have? Draw a picture.

Patterns in Money

NCTM Standards 1, 2, 6, 9, 10
TEKS 2.5A, 2.6A, 2.6C, 2.13A

**At the school store, erasers cost 5¢ and rulers cost 7¢.
What is missing in each price list?**

1.   **Price List for Erasers**

Number of Erasers	1	2		4		6
Total Cost	5¢		15¢			

2.   **Price List for Rulers**

Number of Rulers	1		3	4		
Total Cost	7¢		21¢			

Solve each problem.

3. Billy buys 3 erasers. How much
does that cost?

_____¢

He gives the clerk 25¢. How much
change does he get?

_____¢

4. Billy buys 2 erasers and 1 ruler. How
much does that cost? _____¢
What coins can he use to pay the

exact amount? _____

_____

5. Billy gives the clerk 25¢. He buys as
many erasers as he can. How many
erasers is that?

_____ erasers

How much change does he get?

_____¢

6. Billy has 25¢. He buys as many
rulers as he can. How many rulers
can he get?

_____ rulers

How much does that cost?

_____¢

NOTE: Your child is learning to create and extend
money patterns to solve problems.

At the school store, pencils cost 8¢ and markers cost 10¢. What is missing in each price list?

7.   **Price List for Pencils**

Number of Pencils	1	2		4	
Total Cost	8¢		24¢		

8.   **Price List for Markers**

Number of Markers	1		3	4	
Total Cost	10¢				50¢

Complete the table.

	Billy Has	Billy Buys	Total Cost	Change
9.	50¢	1 pencil, 3 markers		
10.	50¢	2 pencils, 3 markers		
11.	50¢	4 pencils, _____ marker	42¢	
12.	50¢	_____ pencils, _____ markers		14¢

Problem Solving

13. Billy wants to spend exactly 50¢. He must buy at least one pencil and one marker. What might he buy?

_____

Name _____ Date/Time _____

Place Value in Money

NCTM Standards 1, 2, 6, 8, 9, 10

TEKS 2.3E, 2.13A

 =

100 cents equals 1 dollar.

100¢ $1.00

What is missing?

	Number of Pennies	Cents	Dollars and Cents
1.	320	320¢	$3.20
2.	86		$0.86
3.	173		
4.		298¢	
5.			$5.04
6.	439		
7.		75¢	

What is each amount in dollars and cents?

8.

_____

9.

_____

 NOTE: Your child is learning to write amounts of money in dollar and cents notation. Together, look through supermarket flyers for prices written with a dollar sign and a decimal point.

How can you show each money amount with the fewest dollar bills and coins?

10. $2.36

11. $1.49

12. $0.82

13. $3.15

 14. Is $2.28 closer to $2.00 or $3.00? Tell how you know.

_____

_____

_____

Challenge
Write each missing amount.

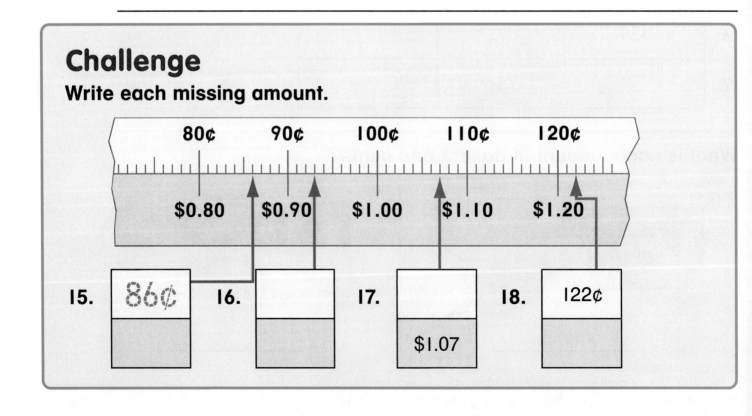

15. 86¢

16.

17. $1.07

18. 122¢

Chapter 10
Lesson 5

Computing with Money

NCTM Standards 1, 2, 6, 8, 9, 10

🔻 TEKS 2.3E

What is missing?

1.

10¢
$0.10

+

4¢
$0.04

=

14¢
$

2.

98¢

+

¢
$0.20

=

3.

$2.53

+

8¢

=

4.

$1.42

+

124¢

=

5.

$2.00 + $4.00 = $ __6.00__

6.

$3.50 + $2.30 = $ ___.___

7.

$5.00 + 45¢ = $ ___.___

8.

$1.20 + 64¢ = $ ___.___

9.

9¢ + $3.82 = $ ___.___

10.

75¢ + $5.50 = $ ___.___

NOTE: Your child is learning to add and subtract money. Give your child a bill and some coins to make up a problem.

Use the menu to solve the problems.

11. Sam buys a salad. He pays $5.00. How much is his change?

    $ _1.55_

Menu	
Salad	$3.45
Turkey sandwich ...	$4.70
Milk	85¢
Apple	60¢

12. Sara buys a turkey sandwich. She pays $5.00. How much is her change?

    _____

13. Lee buys an apple. He pays $1.00. How much is his change?

    _____

14. Use the menu to make up your own problem.

    _____

    _____

    _____

Challenge

15. What is missing? Complete the addition table.

+	5¢	25¢	8¢	$1.00	84¢	$
7¢	12¢	32¢		107¢		148¢
	$0.12	$				

Chapter 10
Lesson 6

Adding Two-Digit Numbers

NCTM Standards 1, 2, 6, 7, 8, 9, 10
TEKS 2.1B, 2.3B

What is missing?

1.

50	7	57
	6	36
	13	

➡️

	5	7
+	3	6
	9	

2.

	4	14
	3	83

➡️

	1	4
+	8	3

3.

		62
		29

➡️

	6	2
+	2	9

4.

		28
		65

➡️

	2	8
+	6	5

NOTE: Your child is learning to add numbers
using a Cross Number Puzzle. Ask your child
to explain how to find the sum of 42 + 39.

What is missing?

5.

6.

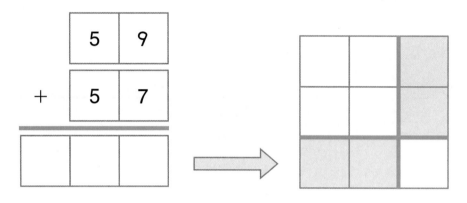

7. What is the sum? Tell how you added these numbers.

$$23 + 14 = \text{_____}$$

_____

_____

_____

_____

Problem Solving

8. There are 19 boys and 17 girls in class. How many children are in class?

_____ children

Show your work.

Chapter 10
Lesson 7

Subtracting Two-Digit Numbers

NCTM Standards 1, 2, 6, 7, 8, 9, 10
TEKS 2.3B

What is missing?

1.

10	50	60
	20	26
	30	

→

6	0
2	6

−

| 3 | |

2.

	11	71
10		12

→

7	1
1	2

−

| | |

3.

4	1
1	4

−

| | |

→

30		41
	4	14

4.

6	2
2	6

−

| | |

→

		62
		26

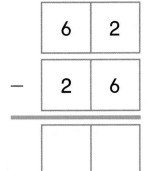

NOTE: Your child is learning how to subtract numbers using
a Cross Number Puzzle. Ask your child to explain how to
find the difference for 57 − 23.

200 + 10 + 13   **CCXXIII**   two hundred twenty-three   **223**

5. What is the difference? Tell how you subtracted these numbers.

$$58 - 26 = \underline{\hspace{2cm}}$$

_____

_____

_____

_____

_____

Write a number sentence to describe each problem. Then find the answers.

6. Lisa scored 19 points. Beth scored 23 points. How many points did both girls score?

_____

_____ points

7. There were 50 balloons at the store. Manny bought 34 of the balloons. How many balloons were still at the store?

_____

_____ balloons

Challenge

8. Mia buys a pencil for 25¢ and an eraser for 35¢. How much change does she get from $1.00?

_____¢

Show your work.

Exploring Expanded Notation

NCTM Standards 1, 2, 6, 7, 8, 9, 10
TEKS 2.1B

What is missing?

1.

$62 = \boxed{\underline{6}0} + \boxed{\underline{}}$

2.

$83 = \boxed{\underline{}0} + \boxed{\underline{}}$

3.

$408 = \boxed{\underline{}00} + \boxed{\underline{}}$

4.

$375 = \boxed{\underline{}00} + \boxed{\underline{}0} + \boxed{\underline{}}$

5.

$981 = \boxed{\underline{}00} + \boxed{\underline{}0} + \boxed{\underline{}}$

6.

400		8	438
	20	0	120
		8	

$\Rightarrow$

4	3	8
1	2	0

+

7.

		4	384
500	0		507

$\Rightarrow$

3	8	4
5	0	7

+

NOTE: Your child is learning to add and subtract three-digit numbers using expanded notation.

What is each sum or difference?

8.

9.

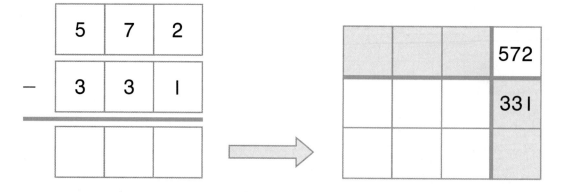

10.

5	7	2
− 3	3	1

Challenge

11. What subtraction does the puzzle show?

200		5	
	60		
	80		180

Mental Math with Three-Digit Numbers

NCTM Standards 1, 6, 7, 8, 9

TEKS 2.1B

Name _____ Date/Time _____

Add 138 in pieces to each number! Do all the work in your head. Write only the answers.

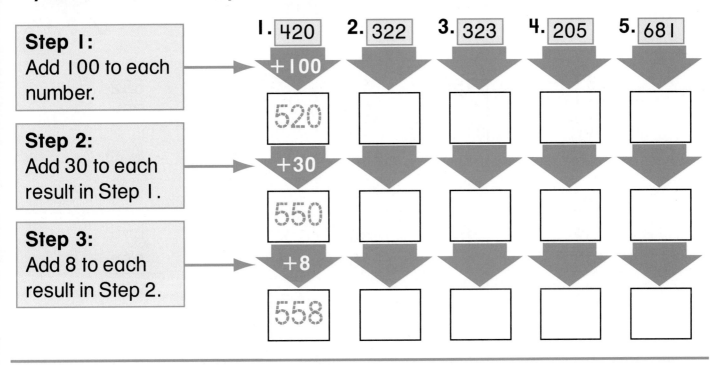

Step 1: Add 100 to each number.

Step 2: Add 30 to each result in Step 1.

Step 3: Add 8 to each result in Step 2.

	1. 420	2. 322	3. 323	4. 205	5. 681
+100	520				
+30	550				
+8	558				

Subtract 254 from any number! Do all the work in your head. Write only the answers.

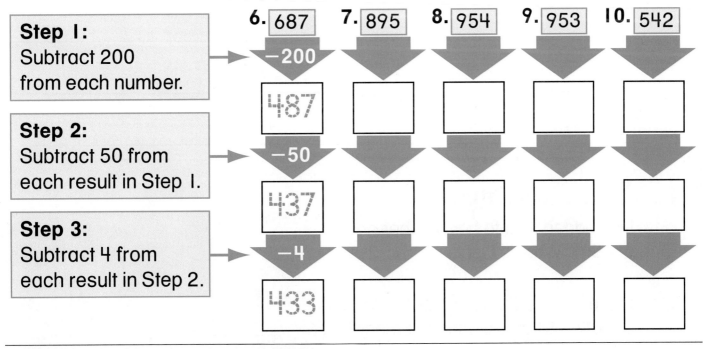

Step 1: Subtract 200 from each number.

Step 2: Subtract 50 from each result in Step 1.

Step 3: Subtract 4 from each result in Step 2.

	6. 687	7. 895	8. 954	9. 953	10. 542
−200	487				
−50	437				
−4	433				

NOTE: Your child is learning to add and subtract numbers using mental math. Ask your child to tell how to add 245 + 160 without a pencil and paper.

Add or subtract.

Think through all of the calculations in your head.

11.   214
     +138
	12. 611
	     +138
		13. 180
		     +138

14.   782
     −254
	15. 682
	     −254
		16. 582
		     −254

17.   452
     +238
	18. 806
	     −154
		19. 474
		     −354

 20. Write an addition or subtraction story using a problem on this page.

_____

_____

_____

Problem Solving

21. Lisa added 200 to her number and got 541. What is Lisa's number?   _____

22. Conor added 50 to his number and got 684. What is Conor's number?   _____

Chapter 10
Lesson 10

Adding Two- and Three-Digit Numbers

NCTM Standards 1, 2, 6, 8, 9, 10
➡ TEKS 2.3B, 2.12A, 2.12B

What is each sum?

1.

2	1	4
+ 7	3	8
		2

200	10	4	214
			738

2.

1	0	5
+ 5	0	7

3.	123 +321	4.	241 +517	5.	146 + 32	6.	319 +435
7.	428 +537	**8.**	842 + 73	**9.**	408 +337	**10.**	318 +152

🏠 **NOTE:** Your child is practicing addition.
Ask your child to add this page number
to the next page number and tell you the sum.

11. Which two numbers in the box have a sum of about 400?

127	976
325	438
509	42

_____ and _____

12. Which two numbers in the box have a sum of about 900?

_____ and _____

Write a number sentence to solve each problem.

13. There are 127 children in the first grade and 146 children in the second grade. How many children are in the two grades?

_____ + _____ = _____   _____ children

14. Shaun read a book with 213 pages. He read another book with 254 pages. How many pages did Shaun read in both books?

_____ + _____ = _____   _____ pages

Challenge

What is each sum?

15.
```
  132
  127
+ 203
-----
```

16.
```
  215
  120
+  64
-----
```

Subtracting Two- and Three-Digit Numbers

NCTM Standards 1, 2, 6, 9, 10
TEKS 2.3B, 2.12A, 2.12B

What is each difference?

1.

5	6	3
− 4	2	7

500	50	13	563

2.

4	8	7
− 1	2	5

3.

6	4	9
−	8	7

4.

2	3	8
− 1	5	4

NOTE: Your child is practicing subtraction. Ask
your child to subtract this page number from 480.

77 + 77 + 77   △   **CCXXXI**   two hundred thirty-one **231**

What is each difference?

5. 789 −310 **479**	6. 645 −222	7. 593 −348	8. 375 − 59
9. 428 −163	10. 247 − 83	11. 708 −276	12. 417 −352

Write a number sentence to solve each problem.

13. There are 863 children at Parkside Elementary School.
There are 124 children in the second grade. How many
children are NOT in the second grade?

_____ − _____ = _____   _____ children

14. Loni is reading a book with 257 pages. She has
already read 36 pages. How many pages does
Loni have left to read?

_____ − _____ = _____   _____ pages

Challenge
What is each missing number?

15.

```
    □
−  352
───────
  176
```

16.

```
  685
−  □
───────
  243
```

Chapter 10
Lesson 12 # Practice Adding and Subtracting

NCTM Standards 1, 2, 6, 8, 9, 10
TEKS 2.3B, 2.13A, 2.14

What is each sum or difference?

I.
573
+321
894

2.
605
+143

3.
784
−412

4.
496
− 73

5.
231
+ 45

6.
591
−245

7.
427
+316

8.
829
−356

9.
362
+418

10.
605
−143

II.
372
+ 45

12.
427
−356

13. Use each of the digits 3, 2, and 5 once.

Make the biggest number. _____

Make the smallest number. _____

Find the sum.

Find the difference.

14. What is the sum? Describe how you added the numbers.

358
+426

_____

_____

_____

_____

15. What is the difference? Describe how you subtracted the numbers.

643
−217

_____

_____

_____

_____

Problem Solving

**Would you add or subtract to solve the problem?
Circle the correct operation.**

16. Harry plays soccer for 45 minutes on Sunday and for 90 minutes on Monday. How long does he play soccer in all?

add   subtract

17. Lorri buys a notebook for $1.99. She pays the clerk with a $5.00 bill. How much change does she get?

add   subtract

Chapter 10
Lesson **13**

Problem Solving Strategy
Solve a Simpler Problem

NCTM Standards 1, 2, 3, 4, 6, 7, 8, 9, 10
⬇ TEKS 2.12A, 2.12B, 2.12C

Understand
Plan
Solve
Check

1. Maria bought 2 erasers. Each eraser
costs 48¢. How much did Maria spend
for the erasers?   _____¢

How did you find the answer?

_____

_____

_____

2. Gabe goes to the library every 4 days. He went on
the 4th day of the year. Will he go to the library on
the 365th day of the year?   _____

How did you find the answer?

_____

_____

_____

3. Joe delivers 60 newspapers every day.
On Sundays he delivers an extra 32 papers. How
many newspapers does Joe deliver in one week?   _____ newspapers

How did you find the answer?

_____

_____

NOTE: Your child is exploring different ways to solve
problems. Sometimes solving a simpler
problem is an efficient way to solve a problem.

Problem Solving Test Prep

1. The first three doors in a hallway are numbered 143, 145, and 147. What is the number of the fifth door?

   Ⓐ 141

   Ⓑ 149

   Ⓒ 150

   Ⓓ 151

2. The movie ended at 8:40. It was 1 hour and 40 minutes long. What time did the movie start?

   Ⓐ 6:00

   Ⓑ 7:00

   Ⓒ 7:20

   Ⓓ 10:20

✏ Show What You Know

3. Carl wants to cut a square into 4 congruent pieces. Name two figures he could make.

   _____

   _____

   _____

   Explain how you found the answer.

   _____

   _____

   _____

4. If 2 people share some cookies, they each get an odd number of cookies. If 3 people share the cookies, they each get an even number of cookies. How many cookies could there be?

   _____ cookies

   Explain how you know.

   _____

   _____

   _____

   _____

   _____

   _____

Chapter 10 Review/Assessment
NCTM Standards 1, 2, 6, 8, 9, 10

What number is missing? Lesson 1

1.

$$85 + \boxed{}$$
_____
100

2.

$$51 + \boxed{}$$
_____
100

What is the value of each collection of coins? Lesson 2

3.

_____¢

4.

_____¢

5. Each sticker costs 6¢. What is missing in the price list? Lesson 3

Price List for Stickers

Number of Stickers	1	2		4		6
Total Cost	6¢		18¢			

6. What is the amount in dollars and cents? Lesson 4

_____

7. What is the sum? Lesson 5

$1.25 + $2.30 = $_____

What is missing? Lessons 6, 7, 8

8.

9.

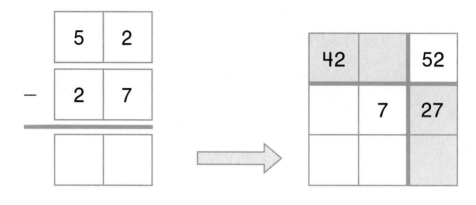

10. $461 =$ [_____00] $+$ [_____0] $+$ [_____]

Find the sum or difference. Lessons 9, 10, 11, 12

| **11.** | 239
 + 29 | **12.** | 560
 +317 | **13.** | 756
 −545 | **14.** | 900
 −730 |

Problem Solving Lesson 13

15. Sean gives the cashier $8.00. He gets
$6.10 in change. How much does he spend?

$_____

Chapter 11

Skip-Counting and Equivalent Sets
Making a Quilt ✏️

🔺 TEKS 2.6C

You need
- 4 squares of paper or cloth
- regular or fabric markers

Make a quilt by arranging squares in a pattern.

STEP 1 Making Rectangles

Arrange your squares to make a rectangle.
How many different rectangles can
you make with all four squares?   _____ rectangles

Draw the different rectangles.

STEP 2 Making a Group Quilt

Combine your squares with the others in your group.
Make a big quilt in the shape of a rectangle. Draw the quilt.

STEP 3 Making Quilt Patterns

Decorate the quilt with a pattern.
Describe your pattern.

_____

_____

_____

_____

Investigation

Dear Family,

Today we started Chapter 11 of *Think Math!* In this chapter, I will explore how to multiply by combining equivalent sets and how to divide by making fair shares. There are NOTES on the Lesson Activity Book pages to explain what I am learning every day.

Here are some activities for us to do together at home. These activities will help me understand multiplication and division.

Love,

_____

Family Fun

How Many?

Work with your child to play this game. Your child will play a similar game in Lesson 2.

- You will need a recording sheet like the one shown below, a number cube, and pennies or other small items like buttons or cereal pieces.

How many items are in each set?	How many sets are there?	How many items are there in all?

- You and your child take turns. For each turn, toss a number cube two times. The first toss shows how many items to put in a set. The second toss shows how many sets to make.
- Find the total number of items.
- Play until you and your child each take 5 turns.

Sharing Cookies

Work with your child to share amounts of cookies fairly.

- You will need 3 plates and a handful of cookies or other small food items.
- Count out any number of cookies into a pile.
- Together, see if the cookies can be shared fairly among 3 people by placing the cookies on the plates.

- Could you share the cookies fairly? Try other amounts of cookies and see which amounts can be shared fairly and which cannot.

Looking For Patterns in Jumps

NCTM Standards 1, 2, 6, 8, 9, 10

TEKS 2.5A

Skip-count on the number lines. Label your jumps.

1.

2.

3.

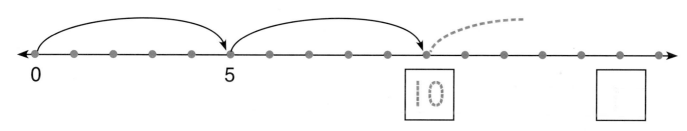

What numbers come next in each pattern?

	1st	2nd	3rd	4th	5th	6th	7th	8th	9th
4. ↻2↗	2	4	6						
5. ↻3↗	3	6	9						
6. ↻5↗	5	10	15						

NOTE: Your child is working more with
skip-counting. Together, look for patterns
when skip-counting by fives.

Skip-count on the number lines. Label your jumps.

7.

| 30 | | | | | |

8.

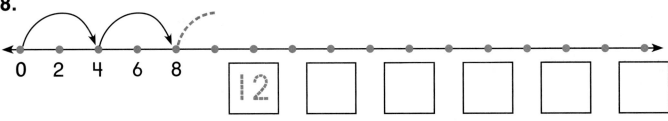

| 12 | | | | | |

9.

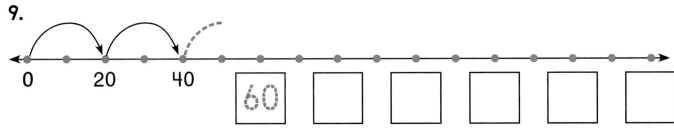

| 60 | | | | | |

What numbers come next in each pattern?

10. ↷10	10	20	30					
11. ↷4	4	8	12					
12. ↷20	20	40	60					

Problem Solving

13. Mike gets 1 nickel each day. How much money will he have in 7 days? Explain. _____¢

_____

_____

Name _____   Date/Time _____

Combining Equivalent Sets

NCTM Standards 1, 2, 6, 7, 8, 9, 10
TEKS 2.4A, 2.13A

How many are there in all?

This is 4 sets of 3, or four threes.

1.

   _12_____ counters

2.

   _____ counters

3.

   _____ cubes

4.

   _____ cubes

5. Draw equivalent sets. Find how many in all.

NOTE: Your child is learning to combine equivalent sets of objects and find the total.

Draw sets of circles. How many are there in all?

6. 3 sets of 6

_____ in all

7. 4 sets of 5

_____ in all

8. Make your own.

_____ sets of _____

_____ in all

Problem Solving

9. Luke buys 3 bags of carrots with 5 carrots in each bag. Molly buys 2 bags of carrots with 10 carrots in each bag. Who buys more carrots?

_____

Use words, numbers, or pictures to explain.

Chapter 11
Lesson 3

Organizing Equivalent Sets

NCTM Standards 1, 2, 6, 7, 8, 9, 10
TEKS 2.4A, 2.13A, 2.14

There will be 4 children at Lynn's party. Each child will get 9 cookies. How many cookies does Lynn need to make?

1. Put the cookies in rows so they are easier to count. Draw them on the cookie sheet. Write how many.

4
_____ rows of _____ cookies

_____ cookies in all

---

2. Now draw the cookies in this grid. Write how many.

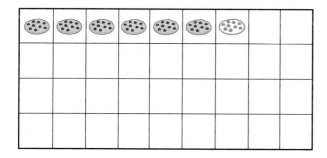

4
_____ rows and _____ columns

_____ cookies in all

NOTE: Your child is learning to arrange objects in an array to make counting easier.

What is missing?

3.

Number of rows	Number in each row	Total
3	4	12

4.

Number of rows	Number in each row	Total
2		

5.

Number of rows	Number of columns	Total
	5	

6.

Number of rows	Number of columns	Total

7.

Number of rows	Number of columns	Total
	3	

8.

Number of rows	Number of columns	Total

Problem Solving

9. Bob has 3 rows of 8 chairs. How else could he put all of the chairs in equal rows? Use words, numbers, or pictures to explain.

Adding Equivalent Sets

NCTM Standards 1, 2, 5, 6, 7, 8, 9, 10
TEKS 2.3A, 2.11B, 2.13A

How many are there in all? Write an addition sentence.

1.

_4_ + _4_ + _4_ = _12_

2.

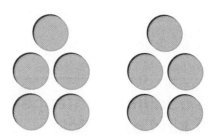

____ + ____ = _____

3.

____ + ____ + ____ = _____

4.

_____

5.

_____

6.

_____

NOTE: Your child is learning to write
addition sentences to add equivalent sets.

The town keeps track of how many vehicles use their bridge every day.

Kind of Vehicle	Vehicles Using the Bridge Today							
cars	😊	😊	😊	😊	😊	😊	😊	
trucks	😊	😊	😊					
vans	😊	😊	😊					
buses	😊	😊						

Key: Each 😊 stands for 8 vehicles.

Write an addition sentence to find the total for each kind of vehicle.

7. cars _____ _____ cars

8. trucks _____ _____ trucks

9. vans _____ _____ vans

10. buses _____ _____ buses

---

 11. Write your own question about the pictograph. Show how to solve the problem.

_____

_____

Problem Solving

12. Alex wrote 3 + 3 + 3 + 3 + 3 = 15 for an array. What other number sentence could he write for the same array?

_____

Draw a picture to explain.

Chapter 11
Lesson 5

Working with Rectangular Arrays

NCTM Standards 1, 2, 6, 7, 8, 9, 10
TEKS 2.4A, 2.13A, 2.14

What is missing?

1.

Rows	Columns	Squares
2	9	

_____ 2 _____ × _____ 9 _____ = _____

2.

Rows	Columns	Squares

_____ × _____ = _____

3.

Rows	Columns	Squares

_____ × _____ = _____

4.

Rows	Columns	Squares

_____ × _____ = _____

5.

Rows	Columns	Squares

_____ × _____ = _____

6.

Rows	Columns	Squares

_____ × _____ = _____

NOTE: Your child is learning to write multiplication sentences for arrays.

Write one addition sentence and one multiplication sentence for each array.

7.

_3_ × _5_ = _____

_5_ + _5_ + _5_ = _____

_____

8.

_____ × _____ = _____

_____

9. Make your own array.

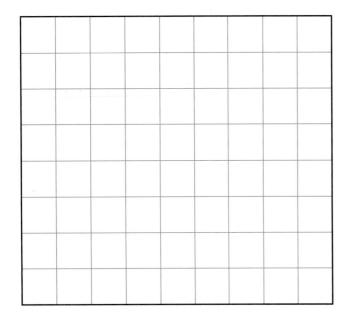

_____ × _____ = _____

_____

Challenge

10. Write two different multiplication sentences for this array.

_____ × _____ = _____   _____ × _____ = _____

Chapter 11
Lesson 6

Building Multiples

NCTM Standards 1, 2, 6, 8, 9, 10
TEKS 2.4A, 2.5A

How many are there?

1. How many eggs are in the box?

6 eggs in each row

2 rows

$2 \times 6 =$ _____ $6 \times 2 =$ _____

_____ eggs in all

2. How many cookies are on the tray?

6 cookies in each row

3 rows

$3 \times 6 =$ _____ $6 \times 3 =$ _____

_____ cookies in all

3. How many legs are on 3 dogs?

3 dogs

4 legs on each dog

$4 \times 3 =$ _____ $3 \times 4 =$ _____

_____ legs in all

4. How many flowers are in 2 vases?

2 vases

3 flowers in each vase

$2 \times 3 =$ _____ $3 \times 2 =$ _____

_____ flowers in all

5. How many fingers are on 4 hands?

4 hands

5 fingers on each hand

$4 \times 5 =$ _____ $5 \times 4 =$ _____

_____ fingers in all

6. How many days are in 4 weeks?

7 days in each week

4 weeks

$7 \times 4 =$ _____ $4 \times 7 =$ _____

_____ days in all

NOTE: Your child is learning to find multiples of a number. Together find how many shoes are in the house.

Complete each table.

7. How many wheels are on 9 tricycles?

Number of Tricycles	1	2	3	4	5	6	7	8	9
Number of Wheels	3	6							

8. How many legs are on 9 chairs?

Number of Chairs	1	2	3	4	5	6	7	8	9
Number of Legs	4	8							

9. A spider has 8 legs. How many legs do 9 spiders have? Make a table to find out. _____ legs

Problem Solving

10. Steve has 6 weeks to finish a project. There are 7 days in a week. How many days does Steve have to finish the project? _____ days

Sharing Between Two Children

NCTM Standards 1, 2, 6, 7, 8, 9, 10
TEKS 2.4B

How many does each child get? Draw to share each amount equally between 2 children.

Try it with counters.

1. 16 cookies

Each child gets ___8___ cookies.

2. 10 cookies

Each child gets _____ cookies.

3. 22 cookies

Each child gets _____ cookies.

4. Make your own. _____ cookies

Each child gets _____ cookies.

NOTE: Your child is learning to divide amounts into two equivalent sets.

Share each amount in 2 equivalent sets.
Use a different color for each set.

5. ◯◯◯◯◯◯◯◯◯◯◯◯◯◯

| 14 balls | ⟹ | 2 shares | ⟹ | _7_ balls each |

6. ♡♡♡♡♡♡♡♡♡♡♡♡♡♡♡♡♡♡♡♡

| 20 hearts | ⟹ | 2 shares | ⟹ | _____ hearts each |

7. ▱▱▱▱▱▱▱▱▱▱▱▱▱▱▱▱▱▱

| 18 blocks | ⟹ | 2 shares | ⟹ | _____ blocks each |

8. Draw any number of objects. Try to make 2 equivalent sets. Explain what you did.

_____

_____

_____

Problem Solving

9. I have 8 of the same coins. I share the coins equally in 2 pockets. How much money might be in each pocket?

Show how you solved the problem.

_____ ¢

Chapter 11
Lesson 8

Sharing Among Three Children

NCTM Standards 1, 2, 6, 7, 8, 9, 10
TEKS 2.4B

How many does each child get? Draw to share each amount equally among 3 children.

1. 12 cookies

Each child gets ___4___ cookies.

2. 21 cookies

Each child gets _____ cookies.

3. 30 cookies

Each child gets _____ cookies.

4. Make your own. _____ cookies

Each child gets _____ cookies.

NOTE: Your child is learning to divide amounts equally into three sets. Together, try to divide a bunch of pennies equally into 3 piles.

Share each amount in 3 equivalent sets.
Use a different color for each set.

5. ⬤ ⬤ ⬤ ◯ ◯ ◯ ◯ ◯ ◯

| 9 balls | ➡ | 3 shares | ➡ | 3 balls each |

6. ♡♡♡♡♡♡♡♡♡♡♡♡♡♡♡♡♡♡♡♡♡♡♡♡

| 24 hearts | ➡ | 3 shares | ➡ | _____ hearts each |

7. ▱▱▱▱▱▱▱▱▱▱▱▱▱▱▱

| 15 blocks | ➡ | 3 shares | ➡ | _____ blocks each |

8. Look back at Problems 5 to 7. Which of these sets can also be divided into 2 equivalent sets? Tell how you know.

_____

_____

_____

Challenge

9. What is missing?

3	6	9			18	21	
1	2	3	4				8

Chapter 11
Lesson 9

How Many Packages?

NCTM Standards 1, 2, 6, 8, 9, 10
🔺 **TEKS 2.4B**

**How many packages can you fill?
Complete each order. Use counters
or draw a picture.**

The factory can put any number of wheels in a package for special orders.

1. Start with 15 wheels.
Put 5 in each package.

Fill ___3___ packages.

---

2. Start with 24 wheels.
Put 4 in each package.

Fill _____ packages.

---

3. Start with 27 wheels.
Put 3 in each package.

Fill _____ packages.

---

4. Start with 48 wheels.
Put 6 in each package.

Fill _____ packages.

Make your own.

5. Start with _____ wheels.

Put _____ in each package.

Fill _____ packages.

NOTE: Your child is learning to divide
amounts into equivalent sets and find
how many sets.

How many sets can you make?

6.

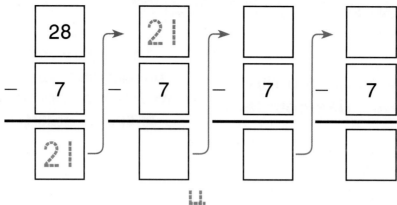

There are ___4___ sets of 7 in 28.

7.

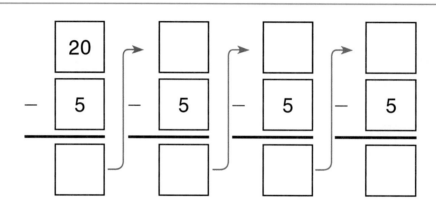

There are _____ sets of 5 in 20.

8.

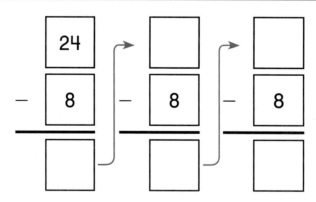

There are _____ sets of 8 in 24.

Problem Solving

9. Larry is packing an order of wheels. He fills
   3 packages of 6 wheels each. He has 12 wheels
   left to pack. How many wheels are in the total order?   _____ wheels

Problem Solving Strategy
Make a List ✏

NCTM Standards 1, 2, 4, 5, 6, 7, 8, 9, 10

TEKS 2.12B, 2.12C

Understand
Plan
Solve
Check

1. Erasers come in packs of 4. Pencils come in packs of 6. I want to buy the same number of erasers and pencils. How many packs of erasers and pencils do I need to buy?

   1. 4 erasers, 6 pencils
   _____
   _____
   _____
   _____

   _____ packs of erasers

   _____ packs of pencils

2. Paula has 3 shirts. One is red, one is yellow, and one is green. She has two skirts. One is black and one is white. How many different outfits can she make?

   _____ outfits

3. Sid uses a toothpick to make each side of a triangle. He makes 5 triangles. How many toothpicks does he need?

   _____ toothpicks

4. Lisa, Max, and Nate are running in a race. They finish first, second, and third. How many different ways can they finish the race?

   _____ ways

NOTE: Your child is exploring different ways to solve problems. Sometimes making a list is an efficient way to solve a problem.

Problem Solving Test Prep

1. At a bus stop, 3 people get on and 2 people get off. Now there are 26 people on the bus. How many people were on the bus before it stopped?

   Ⓐ 5 people

   Ⓑ 25 people

   Ⓒ 26 people

   Ⓓ 28 people

2. A snail travels 1 foot every 5 minutes. He starts crawling at 6:15. What time will it be when he has traveled 10 feet?

   Ⓐ 6:05

   Ⓑ 6:20

   Ⓒ 6:50

   Ⓓ 7:05

Show What You Know

3. David has 36 chairs. Half of the chairs have pads. How many chairs do not have pads?

   _____ chairs

   Explain how you found the answer.

   _____

   _____

   _____

   _____

   _____

   _____

4. Doris has 3 quarters. She wants to buy two notebooks. Each notebook costs 49¢. Does she have enough to buy both notebooks?

   _____

   Explain how you know.

   _____

   _____

   _____

   _____

   _____

   _____

Chapter **11** **Review/Assessment**
NCTM Standards 1, 2, 6, 7, 8, 9, 10

1. Skip count on the number line. Label your jumps. Lesson 1

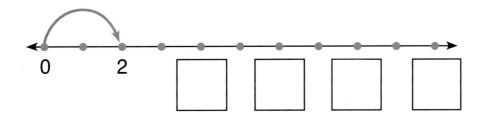

2. How many are there in all? Lesson 2

_____ counters in all

3. What is missing? Lesson 3

Number of rows	Number in each row	Total
2		

How many are there in all? Write an addition sentence. Lesson 4

4.

_____ + _____ + _____ = _____

5.

_____ + _____ = _____

6. Write one addition sentence and one multiplication sentence for the array. Lesson 5

_____

_____ × _____ = _____

7. How many in all? Complete the table. Lesson 6

Number of Hands	1	2	3	4	5	6	7	8	9
Number of Fingers	5	10							

8. How many does each child get? Draw to share 12 cookies equally between 2 children. Lesson 7

Each child gets _____ cookies.

9. Share 6 balls equally among 3 children. Lesson 8

| 6 balls | ⇨ | 3 shares | ⇨ | _____ balls each |

10. How many packages can you fill? Complete the order. Use counters or draw a picture. Lesson 9

Start with 16 wheels.
Put 4 in each package.

Fill _____ packages.

Problem Solving Lesson 10

11. Mary uses a toothpick to make each side of a square. She makes 5 squares. How many toothpicks does she need?

_____ toothpicks

Chapter 12 **Measuring Length**
How Far Is It? ✏️

Try different ways to measure distance.

STEP 1 Estimating Distance

Stand on one side of your classroom. Look across the room. Estimate the distance to the other side of the room.

It looks about _____ long.

STEP 2 Measuring Distance

Talk with your group about how you could measure the length of the room. Then try it. What did you find?

_____

_____

It measures about _____ long.

STEP 3 Comparing Lengths

Compare lengths with other groups. How were the results alike? How were they different?

_____

_____

_____

_____

Investigation

Dear Family,

Today we started Chapter 12 of *Think Math!* In this chapter, I will learn different ways and different units to measure length. There are NOTES on the Lesson Activity Book pages to explain what I am learning every day.

Here are some activities for us to do together at home. These activities will help me as I learn to measure length.

Love,

_____

Family Fun

What's My Length?

Work with your child to play this game. Your child will play this game in Lesson 4.

■ Think of a simple measurement in inches, feet, or yards. For example, you might use 1 foot, 6 inches, or 5 yards. Do not share the measurement with your child. Keep it a secret.

■ Have your child ask *yes/no* questions to try and guess your secret measurement. Some possible questions might be:

  • Is it longer than 1 foot?

  • Is it as long as this room?

  • Is it shorter than my hand?

■ Have your child continue asking questions until he or she has correctly identified the measurement.

■ Switch roles and play again.

Reading Road Signs

Together, find road signs listing different distances.

■ When traveling around town or on a highway, play an informal game to find road signs that give different distances.

■ Call them out as you see them. After reading each one, tell whether it is a short distance or a long distance.

DETOUR 500 FEET

14 FT 4 IN.

■ After returning home, you might want to draw some of the signs you saw on your trip.

Chapter 12
Lesson 1

Measuring Length with Nonstandard Units

NCTM Standards 1, 4, 6, 9, 10

TEKS 2.13A

My group is measuring with _____ as the unit.

Draw a line for each length.

I. I unit long

●

_____

2. 2 units long

●

_____

3. 3 units long

●

_____

4. 4 units long

●

_____

5. Make your own line. Measure it and write the length.

●

about _____ units long

_____

NOTE: Your child is learning about nonstandard units. Have your child measure objects around your home using pennies as a unit of measure.

How long is each bar? Use your unit to measure.

6.

The bar is about _____ units long.

7.

The bar is about _____ units long.

8.

The bar is about _____ units long.

9. Draw a bar. Find its length with your unit.

The bar is about _____ units long.

Challenge

10. How long is the path? Use your unit.

The path is about _____ units long.

Name _____   Date/Time _____

Measuring to the Nearest Inch

NCTM Standards 1, 4, 6, 8, 9, 10
 TEKS 2.9A, 2.12A, 2.13A

How long is the picture of each object? Use a ruler to measure to the nearest inch.

1.

inches   1   2   3   4   5   6

about ____6____ inches

> **Line up the end of the ruler with the end of the object.**

2.

about _____ inches

3.

about _____ inches

4.

about _____ inches

5.

about _____ inch

 NOTE: Your child is learning to measure length to the nearest inch.

How long is each real object? Estimate.
Then measure to the nearest inch.

	Object	Estimate	Measurement
6.	classroom door	about _____ inches	about _____ inches
7.	desk	about _____ inches	about _____ inches
8.	math book	about _____ inches	about _____ inches

Draw a picture of something in your
classroom that matches each length.

9. about 2 inches

10. about 5 inches

11. about 10 inches

Problem Solving

12. Todd is buying a new kite. He wants one that is more than 10 inches wide. He goes to the store without a ruler. How might he measure the kite?

_____

_____

Measuring in Inches, Feet, and Yards

NCTM Standards 1, 4, 6, 8, 9, 10
TEKS 2.9A, 2.12A, 2.13A

Which unit would you use to measure each real object?
Circle *inches*, *feet*, or *yards*.

1.

inches

feet

(yards)

2.

inches

feet

yards

3.

inches

feet

yards

4.

inches

feet

yards

5.

inches

feet

yards

6.

inches

feet

yards

7.

inches

feet

yards

8.

inches

feet

yards

9.

inches

feet

yards

 NOTE: Your child is learning about inches, feet, and yards. Ask your child to name an object or distance that would be best to measure in yards.

Draw a classroom object or distance you would measure with each unit. Find the length.

	Unit	Object	Length to the Nearest Unit
10.	inches		_____ inches
11.	feet		_____ feet
12.	yards		_____ yards

Write *inch, foot,* or *yard* for each benchmark object.

13.

about 1 _____

14.

about 1 _____

15.

a little more than 1

_____

Problem Solving

16. Jan says that she measured her book with string. How could string help her find the width?

_____

_____

Chapter 12
Lesson 4

Relating Inches, Feet, and Yards

NCTM Standards 1, 2, 4, 6, 8, 9, 10
TEKS 2.6A, 2.6C, 2.12A, 2.13A

Which is the best unit to measure the real object?
Write *inches*, *feet*, or *yards*.

1.

about 4 _____ long

2.

about 6 _____ long

3.

about 5 _____ long

4.

about 5 _____ long

5.

about 6 _____ tall

NOTE: Your child is learning how inches, feet, and yards
are related. Ask your child to tell how many inches are in
I foot and how many feet are in I yard.

Complete each table.

6.

Number of Yards	1	2	3		5	6	
Number of Feet	3			12			21

7.

Number of Feet	1	2	3	4	5		7
Number of Inches	12	24				72	

Use the tables to complete the problems.
Then write your own problems.

8. The chalkboard is 2 yards long. How many feet is that?

_____ feet

9. The baseball bat is 3 feet long. How many inches is that?

_____ inches

10. Write a problem about yards and feet.

_____

_____

_____

11. Write a problem about feet and inches.

_____

_____

_____

Challenge

12. Each time George Grasshopper jumps, he moves twice as far as his last jump. His first jump is 3 inches. On which jump will he move 4 feet?

jump _____

Jump Number	1	2	3	4	5	6
Distance in Inches	3	6	12			

Name _____ Date/Time _____

Using Fractions to Measure Length

NCTM Standards 1, 2, 4, 6, 7, 8, 9, 10

TEKS 2.2A, 2.6A, 2.6C, 2.12A, 2.13A

**How long is each piece of ribbon? Use a ruler
to measure to the nearest half inch.**

1.

inches
1   2   3   4   5   6

about ___$4\frac{1}{2}$___ inches

2.

about _____ inches

3.

about _____ inches

4.

about _____ inches

5. Draw a line that is about $3\frac{1}{2}$ inches long.
Start at the dot.

•

NOTE: Your child is learning
about half inches and fractions of
a foot and of a yard.

Complete each table. What is missing?

6.

Number of Yards	$\frac{1}{3}$	$\frac{2}{3}$	1		3	4	
Number of Feet	1	2		6			15

7.

Number of Feet	$\frac{1}{12}$	$\frac{2}{12}$	$\frac{5}{12}$	$\frac{7}{12}$	$\frac{11}{12}$		2
Number of Inches	1	2				12	

Use the tables to complete the problems.

8. Tina cuts a 6-inch piece of ribbon. What fraction of a foot is the piece of ribbon?

_____ of a foot

9. Bob jumps 24 inches. Kyle jumps 1 yard. Who jumped further?

_____

10. Jill has 7 feet of ribbon. Amy has 2 yards of ribbon. Who has more ribbon? Use words, numbers, or pictures to explain.

Problem Solving

11. Can a piece of ribbon be 4 inches long when measured to the nearest inch and to the nearest half inch? Explain.

_____

_____

_____

Chapter 12
Lesson 6

Measuring to the Nearest Centimeter

NCTM Standards 3, 4, 6, 9, 10
TEKS 2.7A, 2.12A, 2.13A

How long is the picture of each object? Use a ruler to measure to the nearest centimeter.

1.

about __14__ centimeters

2.

about _____ centimeters

3.

about _____ centimeters

4.

about _____ centimeters

5.

about _____ centimeter

6.

about _____ centimeters

NOTE: Your child is learning to measure length to the nearest centimeter.

Draw a line for each length. Start at the dot.

7.

•‑ ‑ ‑ ‑

13 centimeters

8.

•

2 centimeters

9.

•

8 centimeters

Draw a picture of something in your classroom that matches each length.

10. about 5 centimeters | **11.** about 30 centimeters | **12.** about 45 centimeters

Challenge

Follow the clues to draw a figure.

13. The distance around a figure is 12 centimeters. The figure has 4 sides. The sides are NOT all the same length.

Chapter 12
Lesson 7 **Measuring in Centimeters and Meters**

NCTM Standards 1, 4, 6, 7, 8, 9, 10
TEKS 2.9A, 2.12A, 2.13A, 2.14

Which unit would you use to measure each real object?
Circle *centimeters* or *meters.*

I.

centimeters

(meters)

2.

centimeters

meters

3.

centimeters

meters

4.

centimeters

meters

5.

centimeters

meters

6.

centimeters

meters

7.

centimeters

meters

8.

centimeters

meters

9.

centimeters

meters

NOTE: Your child is learning about
centimeters and meters.

Choose a unit to measure each real object. Then measure to the nearest unit.

Units of Measure
inches   feet   yards
centimeters   meters

	Object	Estimate	Measurement
10.	chalk	_____	about _____
11.	ABCDEFGHIJ KLMNOPQRST UVWXYZ chalkboard	_____	about _____
12.	book	_____	about _____

13. For Problem 12, how did you choose the unit? How did you measure the book? _____

_____

_____

Problem Solving

14. Kate and Carla each have a piece of string. Kate's string is 7 centimeters long. Carla's string is 7 inches long. Who has the longer string? _____

Use words, numbers, or pictures to explain.

Problem Solving Strategy
Act It Out

NCTM Standards 1, 2, 4, 6, 7, 8, 9, 10
TEKS 2.12A, 2.12B, 2.12C, 2.13A

Understand
Plan
Solve
Check

I. A pencil case is 15 centimeters long.

Can your pencil fit in this case? _____

How long is your pencil? about _____

2. Pat's pencil is longer than this line.

How long might Pat's pencil be?   about _____

3. Tyler buys a book for 86¢. He pays the exact amount with the fewest coins. What coins does Tyler use?

4. Alice has 12 shells, Sandy has 9 shells and Wyatt has 15 shells. They want to share the shells equally. How many shells will each person get?

_____ shells

NOTE: Your child is exploring different ways to solve problems. Sometimes acting it out is an efficient way to solve a problem.

Problem Solving Test Prep

1. When you double this odd number, you get a number between 50 and 55. What is the number?

   Ⓐ 27

   Ⓑ 29

   Ⓒ 31

   Ⓓ 53

2. There are 8 pages in a photo album. Each page has 6 photos on it. How many photos are in the whole album?

   Ⓐ 14 photos

   Ⓑ 24 photos

   Ⓒ 36 photos

   Ⓓ 48 photos

✏ Show What You Know

3. Jan has already delivered 135 newspapers. He has 57 more newspapers to deliver. How many newspapers will Jan deliver in all?

   _____ newspapers

   Explain how you found the answer.

   _____

   _____

   _____

4. Two counters are tossed on this gameboard. Add the two scores. What total scores are possible?

   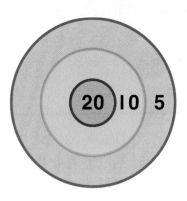

   _____

   Explain how you found the answer.

   _____

   _____

   _____

   _____

Chapter **12** Review/Assessment
NCTM Standards 1, 2, 3, 4, 6, 7, 8, 9, 10

1. Use a paper clip as the measuring unit.
Draw a line 3 units long. Lesson 1

●

2. How long is the picture of the crayon?
Use a ruler to measure to the nearest inch. Lesson 2

about _____ inches

3. Which unit would you use to measure a real swing set?
Circle *inches, feet,* or *yards.* Lesson 3

inches   feet   yards

4. Complete the table. Lesson 4

Number of Yards	1	2	3	4		6	7
Number of Feet		6			15	18	

5. How long is the piece of string? Use a ruler to measure to the nearest half inch. Lesson 5

about _____ inches

---

6. How long is the picture of the ribbon? Use a ruler to measure to the nearest centimeter. Lesson 6

about _____ centimeters

---

7. Which unit would you use to measure the nail? Circle *centimeters* or *meters*. Lesson 7

centimeters   meters

Problem Solving Lesson 8

8. Reese has a piece of ribbon that is 2 feet long. She wants to cut it into 3 equal pieces. How many inches long is each piece?

_____ inches

Name _____

Exploring Multiplication and Division

Choosing Snacks ✏

You need
- milk containers
- juice boxes
- pieces of fruit
- pictures of fruits and drinks (optional)

Make and record different combinations.

STEP 1 Making Combinations

How many fruits do you have? _____

How many drinks do you have? _____

What one fruit did you choose? _____

What one drink did you choose? _____

STEP 2 Recording Combinations

What other combination of a drink and fruit could you make?

_____

Record all of the different combinations.

_____

_____

How many different combinations did you find? _____

Investigation

STEP 3 Finding All Combinations

How do you know you found all the combinations?

_____

_____

Dear Family,

   Today we started Chapter 13 of *Think Math!* In this chapter, I will explore combinations, intersections, and arrays as I learn about multiplication and division. There are NOTES on the Lesson Activity Book pages to explain what I am learning every day.

   Here are some activities for us to do together at home. These activities will help me understand multiplication and division.

Love,

_____

Family Fun

What Will I Wear?

Work with your child to act out one of the activities from class.

■ Use some of your child's clothing to find how many different outfits can be made. Take out 2 pairs of pants or skirts and 2 or 3 tops.

■ Work with your child to make outfits by combining each top with a bottom. Help your child come up with a method to make sure you get all the combinations, such as pairing the first top with every bottom and then doing the same with each top.

■ Together, make a list to keep track of all the different outfits. Count all of the different combinations.

■ If time allows, add another top or bottom to see how many more outfits you can make.

Scavenger Hunt

Work with your child to count equal groups.

■ Look around the house to find objects that are arranged in equal rows and columns. For example, tiles on the floor, pictures on a wall, panels on a door, or paints in a box.

■ Help your child find the total number of objects in a display with equal groups. Talk about how to skip-count by the number of objects in a row or column. To find how many eggs are in a full carton, skip-count by twos or by sixes.

■ Help your child write a multiplication sentence to show each arrangement.

$$2 \times 6 = 12 \text{ or } 6 \times 2 = 12$$

Name _____   Date/Time _____

Counting Combinations

NCTM Standards 1, 2, 6, 7, 8, 9, 10
TEKS 2.12A, 2.13A

How many different outfits can be made each time?

1.

_____ outfits

2.

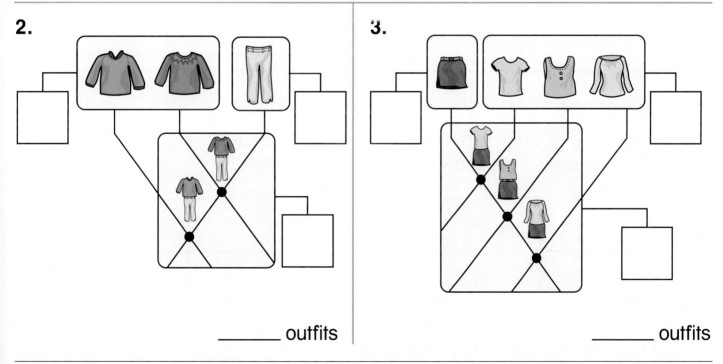

_____ outfits

3.

_____ outfits

4. Andrew has 2 shirts and 2 pairs of shorts for soccer.
How many different uniforms can he make?

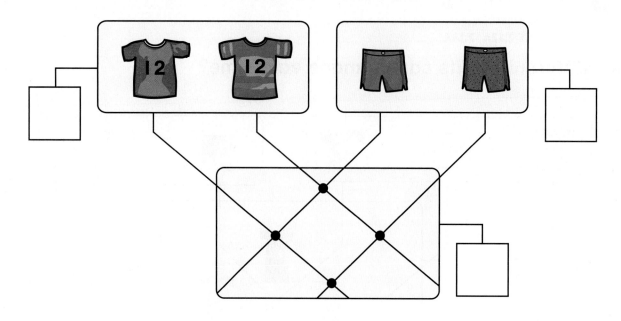

_____ uniforms

What multiplication sentence
can you use to solve the problem?

_____ × _____ = _____

 5. How many different outfits can you make from
3 shirts and 3 pairs of pants? Use words, numbers,
or pictures to explain.

_____ outfits

Challenge

6. How many different two-digit numbers can you make
using the digits 2, 3, and 4? List all of the numbers.

I can make _____ two-digit numbers.

Counting Intersections

NCTM Standards 1, 2, 3, 6, 7, 8, 9, 10

TEKS 2.4A, 2.12A, 2.13A

How many intersections are there? Write the missing numbers.

1.

$5 \times 3 =$ _____

$3 \times 5 =$ _____

_5_____ lines   _____ lines   _____ intersections

2.

$3 \times 4 =$ _____

$4 \times 3 =$ _____

_____ lines   _____ lines   _____ intersections

3. Draw the intersections.

$2 \times 6 =$ _____

$6 \times 2 =$ _____

_____ lines   _____ lines   _____ intersections

NOTE: Your child is working with intersecting lines to learn multiplication facts.

What is missing? Draw lines and numbers to show the multiplication.

4.

$2 \times 4 =$ _____

$4 \times 2 =$ _____

_____ lines   _____ lines   _____ intersections

5.

$3 \times 3 =$ _____

_____ lines   _____ lines   _____ intersections

 6. There are 4 east-west lines and 4 north-south lines. How many intersections are there? Explain how you found the answer.

_____

_____

_____

Problem Solving

7. In Bridgetown, 4 streets go north-south. Every north-south street crosses every east-west street. There is a stoplight at every intersection. There are 24 stoplights in town. How many streets go east-west?

_____ east-west streets

Name _____   Date/Time _____

Counting Hidden Intersections

NCTM Standards 1, 2, 3, 6, 8, 9, 10
TEKS 2.4A, 2.12A, 2.13A

How many intersections are there?
Write the missing numbers.

I.

$2 \times 1 =$ _2_

$1 \times 2 =$ _2_

2.

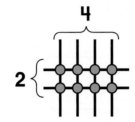

$2 \times 4 =$ _____

$4 \times 2 =$ _____

3.

$3 \times 2 =$ _____

$2 \times 3 =$ _____

4.

$3 \times 3 =$ _____

5.

$4 \times 3 =$ _____

_____ $\times$ _____ = _____

6.

$4 \times 5 =$ _____

_____ $\times$ _____ = _____

7.

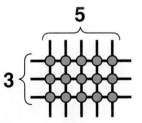

_____ $\times$ _____ = _____

_____ $\times$ _____ = _____

8.

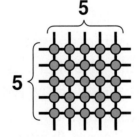

_____ $\times$ _____ = _____

_____ $\times$ _____ = _____

NOTE: Your child is working with intersecting lines to learn multiplication facts.

How many intersections are there?
Write the missing numbers.

9.

5

1

_____ × _____ = _____

_____ × _____ = _____

10.

6

4

_____ × _____ = _____

_____ × _____ = _____

11.

_____

_____

_____ × _____ = _____

_____ × _____ = _____

12.

_____

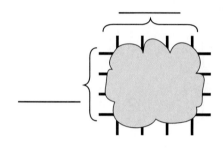

_____

_____ × _____ = _____

13. Choose an even number between 10 and 30.
Draw a town map with that many intersections.

_____ intersections

Problem Solving

14. Bear Town has 8 intersections.
Every north-south street crosses
every east-west street. How many
streets could there be?
Explain how you found the answer.

Draw streets to
make 8 intersections.
Count the streets.

_____ streets

_____

_____

Introducing Division

NCTM Standards 1, 2, 3, 6, 8, 9, 10
TEKS 2.4B, 2.13A

Write the missing numbers.

1.

2.

3.

4.

5.

6.

7.

8.

9.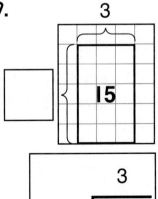

NOTE: Your child is learning to divide by thinking about missing factors.

Write the missing numbers.

10. **11.** **12.**

Here is a shorter way to write the examples for Problems 10 to 12. Write the missing numbers.

13. **14.** **15.**

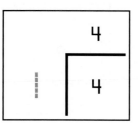

	4
	4

	3
4	

2	8

Challenge

16. What is the missing number?
Tell how you know.

_____

_____

_____

$$8 \begin{cases} 72 \end{cases}$$

Multiplication and Division Fact Families

NCTM Standards 1, 2, 3, 6, 8, 9, 10
TEKS 2.4A, 2.4B, 2.13A

Complete each fact family.

1.

$4 \times 6 = \underline{24}$ $6 \times 4 = \underline{}$

2.

$8 \times \underline{} = 24$ $3 \times \underline{} = 24$

3.

$9 \times 1 = \underline{}$ $1 \times 9 = \underline{}$

4.

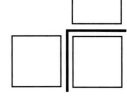

$\underline{} \times \underline{} = \underline{}$ $\underline{} \times \underline{} = \underline{}$

NOTE: Your child is learning to write all members of multiplication and division fact families.

5. Some friends share 8 cookies. Write the missing numbers.
Show the fact family for the pictures.

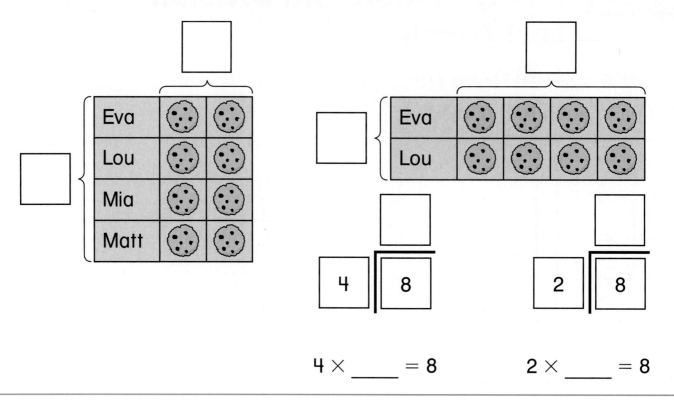

$$4 \times \underline{\qquad} = 8 \qquad\qquad 2 \times \underline{\qquad} = 8$$

6. Draw a rectangle on the grid.
Write the fact family for the rectangle.

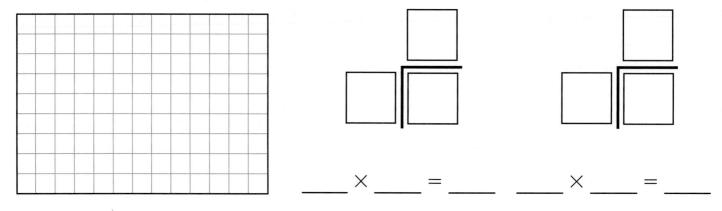

$$\underline{\qquad} \times \underline{\qquad} = \underline{\qquad} \qquad \underline{\qquad} \times \underline{\qquad} = \underline{\qquad}$$

Challenge

7. Write a fact family for a
rectangle with 20 squares

$$\underline{\qquad} \times \underline{\qquad} = \underline{\qquad}$$

$$\underline{\qquad} \times \underline{\qquad} = \underline{\qquad}$$

Chapter 13
Lesson 6

Multiplication and Division Models

NCTM Standards 1, 2, 6, 7, 8, 9, 10
TEKS 2.4A, 2.4B, 2.12A, 2.13A

Complete each model.
What are the missing numbers?

1. $3 \times 5 =$ _____

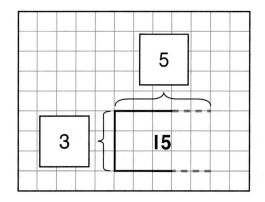

2. $7 \times 3 =$ _____

3. $2 \times 6 =$ _____

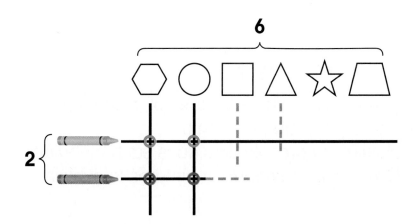

NOTE: Your child is learning basic
multiplication facts by working with
different models.

How can you solve each problem? Show your work.

4. Two children equally share a pack of 18 stickers. How many stickers does each child get?

   _____ stickers

5. Callie is making a sandwich. She has 2 different cheeses and 3 different lunch meats. How many different sandwiches of one meat and one cheese can she make?

   _____ sandwiches

6. Five children go to the fair. Each child wins 5 goldfish. How many goldfish do they win in all?

   _____ goldfish

Problem Solving

7. Three friends share 2 boxes of granola bars.
   There are 9 bars in each box.
   How many granola bars does each friend get?
   Use words, numbers, or pictures to explain.

   _____ bars

Dividing and Estimating with Coins

NCTM Standards 1, 2, 6, 7, 8, 9, 10

TEKS 2.3D, 2.4A, 2.12A, 2.13A

How many coins make one dollar?
Write the missing numbers.

1.

1¢

<u>100</u> pennies = $1.00

_____ × 1 = 100

2.

5¢

_____ nickels = $1.00

_____ × 5 = 100

3.

10¢

_____ dimes = $1.00

_____ × 10 = 100

4.

25¢

_____ quarters = $1.00

_____ × 25 = 100

5.

50¢

_____ half dollars = $1.00

_____ × 50 = 100

6.

100¢

_____ dollar coin = $1.00

_____ × 100 = 100

NOTE: Your child is learning to work with groups of coins
equal to whole dollar amounts. Together, find how many
dimes are equal to $2.00.

Circle the best estimate for each problem.

About means you do not need an exact amount.

7. You have 32 dimes. About how many dollars do you have?

$1.00 $2.00   ($3.00)

8. You have 25 quarters. About how many dollars is that?

$2.00 $6.00 $8.00

9. You have 21 half dollars. About how many dollars is that?

$5.00 $9.00 $10.00

 10. You have $4.00 in one kind of coin. How many coins might you have? Explain.

_____

_____

_____

_____

Problem Solving

11. Jose is saving quarters. He gets one quarter each day. How many days will he need to save for a total of $2.00?

_____ days

Name _____ Date/Time _____

Problem Solving Strategy
Guess and Check ✏

NCTM Standards 1, 2, 3, 6, 7, 8, 9, 10

🔻 TEKS 2.12A, 2.12B, 2.12C, 2.13A

Understand

Plan

Solve

Check

I. My town has 21 intersections. There are 10 roads. How many north-south roads and east-west roads can my town have? Explain how you found the answer.

_____ north-south roads

_____ east-west roads

_____

_____

_____

2. Three children equally share 27 pennies. How many pennies does each child get? Explain.

_____ pennies

_____

_____

_____

3. Jodie has 2 different pair of pants. She can make 14 outfits. How many shirts does Jodie have? Explain.

_____ shirts

_____

_____

_____

NOTE: Your child is exploring different ways to solve problems. Sometimes using the strategy, *guess and check,* is an efficient way to solve a problem.

Problem Solving Test Prep

1. Jeff has a penny, a nickel, a dime, and a quarter. He picks two coins. Which is NOT an amount of money he could have?

   Ⓐ 15¢

   Ⓑ 21¢

   Ⓒ 26¢

   Ⓓ 35¢

2. A snail travels 2 inches every hour. If it starts moving at 3 o'clock, how far would it get by 5 o'clock?

   Ⓐ 2 inches

   Ⓑ 3 inches

   Ⓒ 4 inches

   Ⓓ 5 inches

Show What You Know

3. Bob has a pack of 16 batteries. He puts the same number of batteries in each toy car. He has enough batteries to get 5 cars running. How many batteries go in each car?

   _____ batteries

   Explain your answer.

   _____

   _____

   _____

   _____

   _____

4. Dana has a rectangular piece of cloth. She cuts the cloth with 4 straight lines to get all triangle pieces. How many triangles does she make?

   _____ triangles

   Explain your answer.

   _____

   _____

   _____

   _____

   _____

Name _____ Date/Time _____

1. How many different outfits can you make? Lesson 1

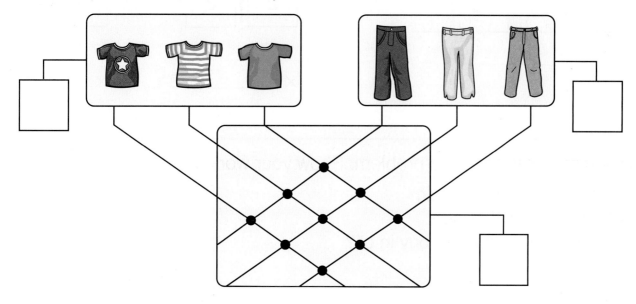

_____ outfits

---

2. What is missing? Draw lines and numbers
to show the multiplication. Lesson 2

$3 \times 4 =$ _____

$4 \times 3 =$ _____

_____ lines   _____ lines   _____ intersections

Write the missing numbers. Lesson 4

3.

4.

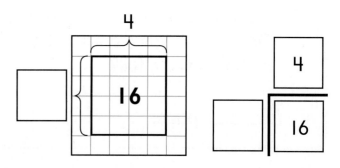

5. Complete the fact family. _{Lesson 5}

  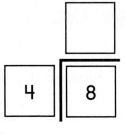

$2 \times$ _____ $= 8$ $4 \times$ _____ $= 8$

---

6. Make a model to solve the problem. Show your work. _{Lesson 6}

Six friends go to a book fair.
Each friend buys 4 books.
How many books do they buy in all?

_____ books

---

7. You have 28 dimes. About how many dollars do you have?
Circle the best estimate. _{Lesson 7}

          $2.00 $3.00 $28.00

Problem Solving _{Lesson 8}

8. Tyson has 20 baseball cards in a book.
One page holds 6 cards. How many pages
of the book are full?

_____ pages

Name _____

Comparing and Contrasting Three-Dimensional Figures

Slide, Stack, and Roll ✏️

You need
- Three-dimensional blocks

What are some ways you can describe a figure?

STEP 1 Exploring a Block

Describe your block.

_____

_____

STEP 2 Sliding, Stacking, and Rolling

Draw each block and put a ✓ if it can slide, stack, or roll.

Figure	Slide	Stack	Roll
▢			
◯			
⬭			

STEP 3 Comparing Blocks

Compare blocks. Describe the blocks that could roll. How are they different from the other blocks?

_____

_____

_____

Investigation

Dear Family,

Today we started Chapter 14 of *Think Math!* In this chapter, I will explore three-dimensional figures and learn how they are alike and different. There are NOTES on the Lesson Activity Book pages to explain what I am learning every day.

Here are some activities for us to do together at home. These activities will help me as I learn about three-dimensional figures.

Love,

_____

Family Fun

What's My Figure?

Work with your child to play this game. Your child will play this game in Lesson 1.

- Think of an object in the house that is shaped like one of the following figures.

cube   rectangular   sphere   prism
prism

For example, a tissue box shaped like a rectangular prism.

- Have your child ask yes/no questions to try and guess your secret object. Some possible questions might be:

  Is there a triangle in the figure?

  Does it roll?

- Have your child continue asking questions until he or she has correctly identified the object.

- Switch roles and play again.

Making Faces

Together, trace a figure to see the faces.

- You will need a three-dimensional object and a pencil and paper.

- Help your child trace around one of the sides of the object.

- Ask your child to name the figure you traced.

- Try other objects. Talk about what sides you can trace and what shapes you will make.

Chapter 14

Lesson 1

Two- and Three-Dimensional Figures

NCTM Standards 2, 3, 6, 8, 9, 10

TEKS 2.7B, 2.13A

Match each object to a set of three-dimensional figures.

1.

2.

3.

4.

5.

6.

7.

spheres

rectangular prisms

cones

cylinders

 NOTE: Your child is learning about three-dimensional figures. Have your child look for objects around your home to match the three-dimensional figures on this page.

Which figure does not belong?

9.

cylinders

10.

pyramids

11.

spheres

12.

rectangular prisms

13.

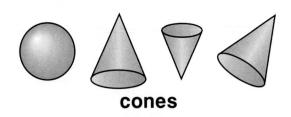

cones

14. Draw an object that does not belong in this set.

pyramids

Name _____ Date/Time _____

Faces

NCTM Standards 2, 3, 6, 7, 8, 9, 10

TEKS 2.7B, 2.13A

Match each three-dimensional figure to its faces.

1.

2.

3.

4.

5.

 NOTE: Your child is learning about the faces of three-dimensional figures.

6. Circle each figure with 6 faces. Put an X on each figure with 5 faces. Put a ☐ on each figure with 0 faces.

rectangular prism   **square pyramid**   **rectangular prism**

triangular prism   **cube**   **triangular pyramid**

rectangular prism   **triangular prism**   **sphere**

---

7. Look at a triangular prism and a triangular pyramid. What is the same about them? What is different?

_____

_____

_____

_____

Challenge

8. What three-dimensional figure has faces in both of these shapes?

_____

Chapter 14
Lesson 3 **Edges**

NCTM Standards 2, 3, 6, 7, 8, 9, 10
TEKS 2.7A, 2.13A

I built these models with toothpicks and marshmallows.

Match each three-dimensional figure to its model. How many toothpicks do you need to make each model?

1.

cube

2.

triangular pyramid

3.

rectangular prism

4.

triangular prism

5.

square pyramid

_____ toothpicks

_____ toothpicks

_____ toothpicks

_____ toothpicks

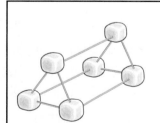

_____ toothpicks

🏠 **NOTE:** Your child is learning about the edges of three-dimensional figures.

How many edges does each object have?

	Object	Number of Edges
6.		
7.		
8.		
9.		

10. Draw your own object.

11. Explain your answer to Problem 9.

_____

_____

_____

Problem Solving

12. The edge of a cube is 2 inches long. How much yarn would you need to cover every edge of the cube? Use words, numbers, or pictures to explain.

_____ of yarn

Name _____ Date/Time _____

Vertices

NCTM Standards 2, 3, 6, 7, 8, 9, 10

TEKS 2.7A, 2.13A

Match each three-dimensional figure to its description.

1.

triangular pyramid

6 faces, 12 edges, 8 vertices

2.

rectangular prism

5 faces, 9 edges, 6 vertices

3.

square pyramid

4 faces, 6 edges, 4 vertices

4.

cube

5.

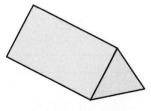

triangular prism

5 faces, 8 edges, 5 vertices

NOTE: Your child is learning about the vertices of three-dimensional figures.

How many faces, edges, and vertices does each object have?

	Object	Number of Faces	Number of Edges	Number of Vertices
6.				
7.				
8.				
9.				
10.	Find your own object.			

Challenge

How many faces, edges, and vertices are in the blocks that make up each building?

11.

_____ faces   _____ edges   _____ vertices

12.

_____ faces   _____ edges   _____ vertices

Chapter 14
Lesson 5

Cylinders and Cones

NCTM Standards 2, 3, 6, 7, 8, 9, 10

TEKS 2.7A, 2.7B, 2.13A

Which figure in each row could have made the footprint?

Footprint	Figures
1.	
2.	
3.	
4.	
5.	
6.	

NOTE: Your child is learning about cylinders and cones. Together find items in the house that look like a cylinder or a cone.

Match each figure to its attribute.

7.

triangular pyramid

a. I vertex

8.

rectangular prism

b. 6 faces

9.

triangular prism

c. 6 edges

10.

cone

d. 5 faces

11.

cylinder

e. 2 faces

Problem Solving

13. What three-dimensional figure can you make from this picture? Tell how you know.

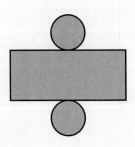

_____

_____

_____

Problem Solving Strategy
Make a Table

NCTM Standards 1, 2, 3, 6, 7, 8, 9, 10

TEKS 2.12A, 2.12B, 2.12C, 2.13A

Understand
Plan
Solve
Check

1. Bob has 5 triangular prisms. He paints the rectangle faces blue and the triangle faces red. How many blue and red faces does he paint?

Number of Prisms	1	2	3	4	5
Rectangle Faces	3				
Triangle Faces	2				10

_____ blue faces   _____ red faces

2. Jana saves $2 this week. Each week she saves double the amount from the week before. How much money will she save after 4 weeks?

_____

Week	Amount Saved
1	$2
2	$4

3. An octopus has 8 arms. How many arms are on 7 octopuses?   _____ arms

Number of Octopuses	1						
Number of Arms	8						

NOTE: Your child is exploring different ways to solve problems. Sometimes making a table is an efficient way to solve a problem.

Problem Solving Test Prep

1. Hope reads every night for 20 minutes. Tonight she finishes at 8:10. What time did she start reading?

   (A) 7:40

   (B) 7:50

   (C) 8:20

   (D) 8:30

2. Piedad starts a number pattern with 3, 7, 11, and 15. If she continues this pattern, what will be the eighth number?

   (A) 17

   (B) 19

   (C) 27

   (D) 31

Show What You Know

3. Jim multiplies a number by itself. The number he gets is between 20 and 30. What is the number?

   _____

   Explain how you found the answer.

   _____

   _____

   _____

   _____

   _____

4. Craig has 12 stickers, Jean has 17 stickers, and Cliff has 13 stickers. They want to share the stickers equally. How many stickers will each person get?

   _____ stickers

   Explain how you found the answer.

   _____

   _____

   _____

   _____

   _____

   _____

Put an X on the figure that does not belong. Lesson 1

1.

2.

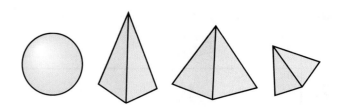

3. Circle each figure with 5 faces.
Put an X on each figure with 6 faces. Lesson 2

4. Look at a cube and a square pyramid.
What is the same about them?
What is different? Lessons 2, 3

_____

_____

_____

_____

_____

How many faces, edges, and vertices does each figure have? Lessons 2–4

	Figure	Number of Faces	Number of Edges	Number of Vertices
5.				
6.				
7.				

Which figure in each row could have made the footprint? Lesson 5

	Footprint	Figures
8.		
9.		

Problem Solving

10. Eric wants to build a square pyramid and a rectangular pyramid from cut-out shapes. How many square faces and triangle faces does he need? Lesson 6

Figure	Square Faces	Triangle Faces
Square Pyramid		
Triangular Pyramid		
Total		

Chapter 15

Capacity, Weight/Mass, and Temperature
Measuring Different Attributes
→ TEKS 2.10A

You need
- coffee mug
- ruler, pan balance, paper cup, water, rice, thermometer

What are all of the different ways that you can measure a mug?

STEP 1 Thinking About Attributes

What can you measure about a mug?

_____

_____

STEP 2 Working with Tools

How might you use tools to measure a mug? _____

_____

_____

_____

STEP 3 Measuring in Different Ways

Use the tools to measure a mug in different ways. What did you find out?

_____

_____

_____

Investigation

Dear Family,

Today we started Chapter 15 of *Think Math!* In this chapter, I will explore how to measure capacity, weight/ mass, and temperature. There are NOTES on the Lesson Activity Book pages to explain what I am learning every day.

Here are some activities for us to do together at home. These activities will help me understand measurement.

Love,

_____

Family Fun

What Is the Temperature?

Work with your child to record the daily air temperature.

- Each day for a week, look in the newspaper to find and record the daily forecast of high and low temperatures.

- Look at your list of temperatures for the week. Talk about which day was the warmest and which was the coldest. Talk about what clothes you might wear in these different temperatures.

- With your child, make a prediction for tomorrow's high and low temperatures.

- Continue recording temperatures beyond the week, if you wish. Then use this information to help your child plan his or her clothes or activities.

Market Measures

Work with your child to identify units of measure at the store.

- Together, look at flyers for the supermarket. Then take a visit to the store and look at different products on the shelves.

- Talk about the weights and liquid measures of meats, juices, cereal, and other products.

- Compare prices of products in different sizes. Decide which is the best buy.

- Have fun shopping and learning together!

Chapter 15
Lesson 1

Comparing, Ordering, and Measuring Capacity

NCTM Standards 2, 4, 6, 7, 8, 9, 10
➡ TEKS 2.9C, 2.12A, 2.13A

Which unit is best to measure the capacity of each container?

spoon   cup   pail

1.

spoons ⟨cups⟩ pails

2.

spoons   cups   pails

Choose the unit that makes the most sense.

3.

spoons   cups   pails

4.

spoons   cups   pails

5.

spoons   cups   pails

6.

spoons   cups   pails

7.

spoons   cups   pails

8.

spoons   cups   pails

NOTE: Your child is learning to compare, order, and measure containers by how much they hold.

9. Becky has a bowl of soup.
Kari has a cup of soup.
Who has more soup?

_____

bowl   **cup**

10. Kate has a pitcher of water. Bud has a
pail of water. Max has a glass of water.
Who has the most water?

_____

pitcher   **pail**   **glass**

11. The metal vase holds 10 cups of water.
The glass vase holds 10 spoonfuls.
Which vase holds less water?

_____

metal vase   **glass vase**

12. How can you find out which of two bowls holds more?

_____

_____

Problem Solving

13. Ray has a red, a blue, and a green pail. The
blue pail holds the most. The green pail holds
more than the red. Which color pail holds
the least? Tell how you found the answer.

_____

_____

_____

Name _____   Date/Time _____

Measuring in Cups, Pints, Quarts, and Gallons

NCTM Standards 1, 2, 4, 6, 8, 9, 10
TEKS 2.12A, 2.13A

Each baby lamb drinks I cup of milk. Match the lambs to what they drink.

> **2 cups = I pint**
> **2 pints = I quart**
> **4 quarts = I gallon**

1.

I pint

2.

3 pints

3.

I cup

4.

2 quarts

5.

3 cups

6.

I quart

 NOTE: Your child is learning about cups, pints, quarts, and gallons. Show your child an empty milk container and ask him or her to find out how many cups will fill the container.

Which measurement from the box solves each riddle?

7. I am more than 1 cup. I am less than 3 cups. What am I?

 I pint
_____

Measurements	
1 pint	3 quarts
1 quart	2 gallons
1 gallon	

8. I am more than 3 cups. I am less than 5 cups. What am I?

_____

9. I am more than 4 pints. I am less than 1 gallon. What am I?

_____

10. I have the same capacity as 16 cups. What am I?

_____

11. I have the same capacity as 8 quarts. What am I?

_____

 12. Make up your own riddle.

_____

_____

_____

Problem Solving

13. Tina has a 1-gallon punch bowl. She wants to fill the bowl with orange juice and lemonade. How much of each can she use?

_____ orange juice   _____ lemonade

Chapter 15
Lesson 3

Measuring in Milliliters and Liters

NCTM Standards 1, 4, 6, 7, 8, 9, 10
TEKS 2.12A, 2.12D, 2.13A

Which unit is better to measure the capacity of each object?

1.

(milliliters)   liters

2.

milliliters   liters

3.

milliliters   liters

4.

milliliters   liters

5.

milliliters   liters

6.

milliliters   liters

Draw something you might measure with each unit.

7.

milliliters

8.

liters

NOTE: Your child is learning about milliliters and liters. Ask your child to guess and then check to find out if various household containers hold more than, less than, or the same as a liter container.

9. Anne drank a glass of milk. Did she drink 230 milliliters or 230 liters of milk?

_____

10. Jeff filled a watering can with water. Did he use 3 liters or 3 milliliters of water?

_____

11. Todd ate a spoonful of soup. Did he eat 15 milliliters or 15 liters of soup?

_____

12. Nina filled the kitchen sink with water. Did she use 25 milliliters or 25 liters of water?

_____

13. Write your own problem. Have a classmate solve it.

_____

_____

_____

Problem Solving

14. The gas tank in Mr. Brown's car holds about 60 liters. Does the gas tank hold more than or less than 60 gallons? Explain how you know. _____

_____

_____

Comparing and Measuring Weight

NCTM Standards 2, 4, 6, 7, 8, 9, 10
TEKS 2.9D, 2.12A, 2.13A, 2.14

Which unit is best to measure the weight of each object?

paper clip   book   brick

I.

(paper clips)   books   bricks

2.

paper clips   books   bricks

3.

paper clips   books   bricks

4.

paper clips   books   bricks

5.

paper clips   books   bricks

6.

paper clips   books   bricks

 7. Why did you choose that unit for Problem 6?
Use words, numbers, or pictures to explain.

NOTE: Your child is learning to compare and measure objects by how much they weigh.

8. Ben's apple weighs more than Ali's apple. Ali's apple weighs more than Casey's apple. Can you tell whose apple weighs the most? Explain.

_____

_____

_____

9. Ali's book weighs more than Casey's book. Ben's book weighs more than Casey's book. Can you tell whose book weighs the most? Explain.

_____

_____

_____

Challenge

10. Label the bags in order from lightest to heaviest. Write A, B, and C.

lightest   heaviest

A   B   B   C

Measuring in Grams and Kilograms

NCTM Standards 1, 2, 4, 6, 8, 9, 10
TEKS 2.12A, 2.13A

What is missing? Complete the table.

	Object	More than or less than 1 gram?	More than or less than 1 kilogram?
1.	marker	more than	
2.	postage stamp		
3.	computer		
4.		more than	less than
5.		more than	more than
6.		less than	

NOTE: Your child is learning to estimate and measure objects in grams and kilograms. Ask your child to name some kitchen objects that are more than 1 kilogram.

Draw two objects from your classroom. Estimate each one in grams or kilograms. Then measure.

	Object	Estimate	Measurement
7.		about _____	about _____
8.		about _____	about _____

9. Find two objects in your classroom that are each about 1 kilogram. Draw them.

Problem Solving

10. An adult cocker spaniel is about 12 kilograms. Would a cocker spaniel puppy measure 3 kilograms or 30 kilograms? Tell how you know.

_____

_____

_____

Chapter 15
Lesson 6 | Measuring in Ounces, Pounds, and Tons

NCTM Standards 1, 4, 6, 9, 10
TEKS 2.12A, 2.13A

Which unit is best to weigh each object?

1.

ounces ⟨pounds⟩ tons

2.

ounces   pounds   tons

3.

ounces   pounds   tons

4.

ounces   pounds   tons

5.

ounces   pounds   tons

6.

ounces   pounds   tons

Draw something you might weigh with each unit.

7.

ounces

8.

pounds

9.

tons

NOTE: Your child is learning about ounces, pounds, and tons. Ask your child to estimate and weigh grocery items.

150 + 150 + 31   **CCCXXXI**   three hundred thirty-one 331

Match each animal to its weight.

10. whale   12 pounds

11. lion   10 ounces

12. cat   125 tons

13. kitten   400 pounds

---

14. Draw a classroom object that you think weighs between 1 pound and 3 pounds. Estimate and measure the weight.

Estimate: about _____ pounds

Measurement: about _____ pounds

Challenge

15. Together, 5 identical marbles weigh 1 ounce. How many marbles would you need to balance a 5-ounce fork?

_____ marbles

Chapter 15
Lesson 7

Measuring Temperature

NCTM Standards 1, 2, 4, 6, 8, 9, 10
TEKS 2.10A, 2.12A, 2.12D, 2.13A

What temperature goes with each picture?
Color the thermometer to show your estimate.

1.

°Fahrenheit

85
80
75
70
65
60
55
50
45
40
35
30
25
20

_____ °F

2.

°Fahrenheit

90
85
80
75
70
65
60
55
50
45
40
35
30
25

_____ °F

3.

°Fahrenheit

90
85
80
75
70
65
60
55
50
45
40
35
30
25
20

_____ °F

4. Draw your own.

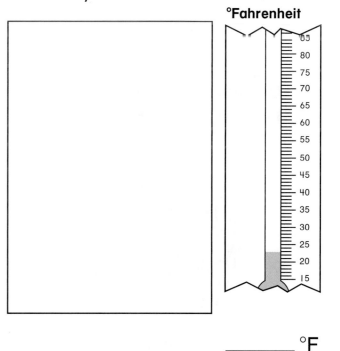

°Fahrenheit

85
80
75
70
65
60
55
50
45
40
35
30
25
20
15

_____ °F

NOTE: Your child is learning how to estimate and measure
temperature in degrees Fahrenheit. Together, read and record the
outside temperature for a week or so using your home thermometer.

Play the cold version of *What's My Temperature?*
Pick a secret temperature between 25°F and −10°F.
Your partner asks *yes/no* questions. Use red and blue markers
to record on the gameboard below.

What's My Temperature? **Gameboard**

Name _____ Date/Time _____

Problem Solving Strategy
Act It Out

NCTM Standards 1, 2, 4, 5, 6, 7, 8, 9, 10

TEKS 2.12A, 2.12B, 2.12C, 2.13A

Understand
Plan
Solve
Check

I. Shayna put 15 cubes in two bags. She put more cubes in Bag B than in Bag A. She balanced the pans by adding 3 cubes to the side with Bag A. How many cubes were in each bag?

Bag A _____ cubes   **Bag B** _____ cubes

2. Pat put 13 cubes in two bags. He put more cubes in Bag A than in Bag B. He balanced the pans by adding 5 cubes to the side with Bag B. How many cubes were in each bag?

Bag A _____ cubes   **Bag B** _____ cubes

3. Deion put 10 cubes in three bags. He put the same number of cubes in Bags A and B. He put more cubes in Bag C than in the other two bags together. He balanced Bags B and C by adding 4 cubes to the side with Bag B. How many cubes were in each bag?

Bag A _____ cubes   **Bag B** _____ cubes   **Bag C** _____ cubes

 NOTE: Your child is exploring different ways to solve problems. Sometimes acting it out is an efficient way to solve a problem.

Problem Solving Test Prep

1. I have 6 faces, 12 edges, and 8 vertices. All of my faces are the same shape. What figure am I?

Ⓐ cube

Ⓑ pyramid

Ⓒ sphere

Ⓓ cone

2. I am thinking of a number. When my number is multiplied by 2 it has a product between 10 and 20. Which is NOT my number?

Ⓐ 9

Ⓑ 8

Ⓒ 6

Ⓓ 4

Show What You Know

3. Monica had 69¢. She bought something at the store. Then she had 45¢ left. How much money did she spend?

_____¢

Explain how you found the answer.

_____

_____

_____

_____

4. Kip just loaded some games onto his computer. He almost doubled the number of games. Now he has 23 games. How many games did he have before?

_____ games

Explain how you found the answer.

_____

_____

_____

_____

Chapter 15 **Review/Assessment**
NCTM Standards 1, 4, 5, 6, 7, 8, 9, 10

Which unit is best to measure the capacity of each container? Lessons 1 and 3

1.

spoons   cups   pails

2.

milliliters   liters

Match each amount to a container. Lesson 2

3. 2 cups

4. 4 cups

5. 16 cups

Which unit is best to weigh each object? Lesson 4

6.

paper   books   bricks
clips

7.

paper   books   bricks
clips

Is each real object *more* or *less* than 1 kilogram? Lesson 5

8.

_____ than 1 kilogram

9.

_____ than 1 kilogram

Match each object to its weight.
Lesson 6

10.   2 tons

11.   2 ounces

12.   2 pounds

13. What temperature goes with the picture? Color the thermometer to show your estimate. Lesson 7

°Fahrenheit

85
80
75
70
65
60
55
50
45
40
35
30

_____ °F

Problem Solving Lesson 8

14. Al put 20 cubes in two bags. He put more cubes in Bag A than in Bag B. He balanced the two bags by adding 4 cubes to the side with Bag B. How many cubes were in each bag?

Bag A _____ cubes   **Bag B** _____ cubes

Chapter 16 Multiplying and Dividing
Recording Different Ways 🖊

TEKS 2.4A

You need
• connecting cubes

STEP 1 Making a Train

As a group, decide how many cubes
to put in a train. Then each person
makes a train of that size.
How many cubes are in your train? _____ cubes

STEP 2 Recording Trains and Cubes

How many trains did your group make? _____ trains
How can you find the number of cubes in all of the trains for your group?

_____

_____

_____

Write a number sentence to show what you did.

_____

STEP 3 Recording a Different Way

Can you think of other number sentences to describe
the trains? Write as many sentences as you can.

_____

_____

_____

Investigation

Dear Family,

Today we started Chapter 16 of *Think Math!* In this chapter, I will learn multiplication and division facts. There are NOTES on the Lesson Activity Book pages to explain what I am learning every day.

Here are some activities for us to do together at home. These activities will help me as I learn to multiply and divide.

Love,

_____

Family Fun

Fact Family Fandango

Work with your child to play this game. Your child will play a similar game in Lesson 3.

■ You and your child each say a number from 0 to 10. Multiply the two numbers and write the sentence on a sheet of paper. For example:

$$6 \times 8 = 48$$

■ Have your child write all the number sentences that are part of the fact family that includes the sentence you just wrote.

$$8 \times 6 = 48, \quad 48 \div 6 = 8, \quad 48 \div 8 = 6$$

■ For every sentence your child records correctly, he or she gets a point. For every sentence he or she misses or records incorrectly, you get a point.

■ Play several rounds. The player with more points at the end of the game wins.

Supermarket Math

Try this activity on your next trip to the store.

■ Look for per-package counts on various products. For example the number of packets of oatmeal in a box, or the number of juice boxes in a pack. Look for packages that contain up to 10 items.

■ Pose simple multiplication and division problems for your child to solve. If you find a 3-pack of juice boxes, then you might ask, "How many juice boxes are in 2 packages? What about 3 packages?"

■ Invite your child to share how he or she found the answer.

Creating Multiplication Tables

NCTM Standards 1, 2, 6, 8, 9, 10
🔹 **TEKS 2.4A, 2.13A, 2.14**

Draw pictures of intersections, sets, or arrays to help you.

1. Fill in the table. Use your memory, draw a picture, or look back at work you have done before.

	0	1	2	3	4	5	6	7	8	9	10
× 0	0		0			0					
× 1											
× 2											
× 3			6		12		18				
× 4											
× 5	0										
× 6											
× 7											
× 8											
× 9											
× 10											

NOTE: Your child is learning to complete a multiplication table using different strategies.

Part of a row from the multiplication table is shown. What number is being multiplied in the row?

2.

× 5		15	20	25	30	35	40

3.

×		28	35	42	49	56	63

4.

×		15	18	21	24	27	30

5. Jan knows that 2 × 9 = 18.
How can she use this fact to solve 3 × 9?

_____

_____

6. How can you use 2 × 9 to solve 9 × 2?

_____

_____

Problem Solving

7. Why is the first row of the multiplication table on page 341 all zeros? Explain.

_____

_____

Name _____ Date/Time _____

Multiplication and Division

NCTM Standards 1, 2, 6, 9, 10

TEKS 2.4A, 2.4B, 2.13A

Each picture is about the numbers 3, 4, and 12.

4 + 4 + 4 = 12

```
  3
  3
  3
+ 3
────
 12
```

What is missing? Complete each fact.

1.

3 × __4__ = 12

2.

4 × _____ = 12

3.

4.

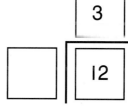

5.

```
    3
  × 4
  ───
  □
```

6.

```
    4
  × 3
  ───
  □
```

7.

_____ ÷ 4 = 3

8.

12 ÷ 3 = _____

NOTE: Your child is learning to complete multiplication and division facts.

How many squares are there in all?
Write the missing numbers.

9.

_____ × _____ = _____

10.

_____ × _____ = _____

11. There are 25 squares in all.
How many rows are there?

25 ÷ _____ = _____

12. There are 24 squares in all.
How many columns are there?

24 ÷ _____ = _____

Challenge
Complete each fact family.

13. $4 \times 5 = 20$

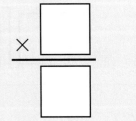

14. $3 \times 6 = 18$

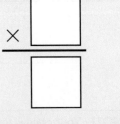

Writing Multiplication and Division Fact Families

NCTM Standards 1, 2, 6, 8, 9, 10

TEKS 2.4A, 2.4B, 2.13A, 2.13B

What is missing? Complete each fact family.

1.

$$3 \times 6 = \boxed{} \qquad \boxed{} \div 6 = \boxed{}$$

$$\boxed{} \times \boxed{} = \boxed{} \qquad \boxed{} \div 3 = \boxed{}$$

2.

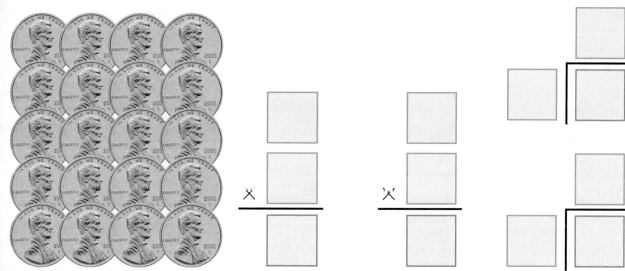

3.

$$4 \times \boxed{} = \boxed{} \qquad \boxed{} \div \boxed{} = \boxed{}$$

$$6 \times \boxed{} = \boxed{} \qquad \boxed{} \div \boxed{} = \boxed{}$$

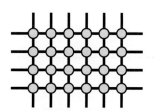

NOTE: Your child is learning to complete multiplication and division fact families.

What is missing? Complete each fact family.

4.

$3 \times \boxed{} = \boxed{}$ $12 \div \boxed{} = \boxed{}$

$\boxed{} \times \boxed{} = \boxed{}$ $\boxed{} \div \boxed{} = \boxed{}$

5.

6.

$3 \times \boxed{} = \boxed{}$ $\boxed{} \div \boxed{} = \boxed{}$

$\boxed{} \times \boxed{} = \boxed{}$ $\boxed{} \div \boxed{} = \boxed{}$

Problem Solving

7. Write a division story. Then write a number sentence to match it.

_____

_____

_____

Chapter 16
Lesson 4

Connecting Pictures, Number Sentences, and Stories

NCTM Standards 1, 2, 6, 8, 9, 10

🔻 **TEKS 2.4A, 2.4B, 2.12A, 2.13A**

Draw a line from each picture to the matching number sentence. Complete the sentence.

1.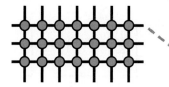

$$28 \div 7 = \underline{\hspace{2cm}}$$

2.

$$7 \times 3 = \underline{\hspace{2cm}}$$

3.

$$32 \div 4 = \underline{\hspace{2cm}}$$

4.

Sun	Mon	Tue	Wed	Thur	Fri	Sat
1	2	3	4	5	6	7
8	9	10	11	12	13	14
15	16	17	18	19	20	21
22	23	24	25	26	27	28

$$16 \div 4 = \underline{\hspace{2cm}}$$

5.

$$4 \times 10 = \underline{\hspace{2cm}}$$

🏠 **NOTE:** Your child is learning to write multiplication and division sentences for pictures.

Complete the fact families to match the pictures and stories.

6.

---

7. There are 24 marbles. They are shared equally among 3 children.

$$\boxed{} \div \boxed{} = \boxed{}$$

$$\boxed{} \div \boxed{} = \boxed{}$$

$$\boxed{} \times \boxed{} \qquad \boxed{} \times \boxed{}$$

---

8. There are 36 baseballs. They are put into packages of 6 baseballs each.

$$\boxed{} \times \boxed{} = \boxed{}$$

Challenge
Find each product.

9. $2 \times 3 \times 4 = $ _____

10. $6 \times 1 \times 5 = $ _____

Name _____   Date/Time _____

Problem Solving Strategy
Act It Out

NCTM Standards 1, 2, 4, 6, 7, 8, 9, 10

TEKS 2.12A, 2.12B, 2.12C, 2.12D, 2.13A

Understand

Plan

Solve

Check

1. Three children equally share $1.50 in quarters.
   How much money does each child get?

   _____ ¢

2. A sheet of stamps has 6 rows with 5 stamps in
   each row. How many stamps are there in all?

   _____ stamps

3. 15 chairs are arranged in equal rows.
   How many chairs might be in each row?

   _____ chairs

4. Some children equally share 48 pretzels. Each
   child gets 8 pretzels. How many children share
   the pretzels?

   _____ children

NOTE: Your child is exploring different ways to solve
problems. Sometimes acting it out is an efficient way to
solve a problem.

Problem Solving Test Prep

1. Cam has 4 coins in his pocket. He has only dimes and pennies. Which total amount is NOT possible for him to have?

   (A) 13¢

   (B) 21¢

   (C) 22¢

   (D) 31¢

2. There are 7 days in a week. Which number sentence shows how many days there are in 3 weeks?

   (A) $7 - 3 = 4$

   (B) $7 + 7 = 14$

   (C) $7 \times 3 = 21$

   (D) $4 \times 7 = 28$

Show What You Know

3. There are 49 erasers packed in boxes. The number of boxes is the same as the number of erasers in each box. How many erasers are in each box?

   _____ erasers

   Explain how you found the answer.

   _____

   _____

   _____

   _____

4. Each wagon has 4 wheels. How many wheels are there on 5 wagons?

Wagon	1				
Wheels	4				

   Explain how you found the answer.

   _____

   _____

   _____

   _____

Chapter 16 Review/Assessment

NCTM Standards 1, 2, 6, 8, 9, 10

1. Complete the multiplication table. Lesson 1

	0	1	2	3	4	5	6	7	8	9	10
× 3			6			15					
× 5					20						
× 7				21							

2. Sue knows that $8 \times 5 = 40$. How can she use
this fact to solve 9×5? Lesson 1

_____

_____

What is missing? Complete each fact. Lesson 2

3.

$7 \times \underline{\hspace{1cm}} = 28$

4.

$\underline{\hspace{1cm}} \div 7 = 4$

5.

$$\begin{array}{r} 7 \\ \times 4 \\ \hline \boxed{} \end{array}$$

6.

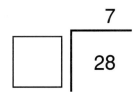

$\boxed{}\,)\,\overline{28}$ with 7

How many squares are there in all? Write the missing numbers. Lesson 2

7.

_____ × _____ = _____

8.

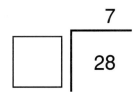

_____ × _____ = _____

9. What is missing? Complete the fact family. Lesson 3

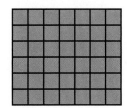

$$6 \times \boxed{} = \boxed{} \qquad 42 \div \boxed{} = \boxed{}$$

$$\boxed{} \times \boxed{} = \boxed{} \qquad \boxed{} \div \boxed{} = \boxed{}$$

Complete the fact families to match the stories. Lesson 4

10. There are 14 balloons. They are shared equally among 7 children.

$$\boxed{} \div \boxed{} = \boxed{}$$

$$\boxed{} \div \boxed{} = \boxed{}$$

$$\times \frac{\boxed{}}{\boxed{}} \qquad \times \frac{\boxed{}}{\boxed{}}$$

---

11. There are 81 books. They are put on shelves with 9 books on each shelf.

$$\boxed{} \times \boxed{} = \boxed{}$$

$$\boxed{} \overline{\big)\boxed{}}$$

Problem Solving Lesson 5

12. Five children equally share $1.00 in dimes.
How much money does each child get?

_____ ¢

10 less

55, 45, 35, 25, 15, 5

Each number is **10 less** than the number before.

10 more

8, 18, 28, 38, 48, 58, 68

Each number is **10 more** than the number before.

add

To join 2 groups

addend

4 + 4 = 8

addends

addition

4 + 1 = 5

after

16 17 18 19 20 21 22 23 24 **25**

25 is **after** 24.

always

Green will **always** be picked.

area

The number of square units that cover a flat surface

area = 10 square units

array

A rectangular arrangement of objects in rows and columns

3 × 4 = 12

bar graph

Hours of Playing Sports Last Week

base

base

before

16 17 18 19 20 21 22 23 24 25

17 is **before** 18.

benchmark(s)

about 1 inch

between

25 26 27 28 29 **30** 31 32 33 34

30 is **between** 29 and 31.

biggest

115   15   5

115 is the **biggest** number.

bottom

calculator

capacity

centimeter

Used to measure the length of short objects

cent(s)

28 **cents**

certain

Green is a **certain** outcome.

change

| I buy an apple for 50¢. | I pay with $1.00. | I get 50¢ **change.** |

A line graph shows **change** over time.

circle

circular

All of these have a **circular** shape.

close (to)

11 is **close** to 10.

closed figures

354

closer (to)

22 is **closer to** 20 than to 30.

coin(s)

column(s)

column ↑

combination

A **combination** of *i* and *n* is *in*.

compare

When you **compare** the two groups, you see that 29 is less than 63.

cone

congruent

Figures that are the same size and shape are **congruent.**

corner(s)

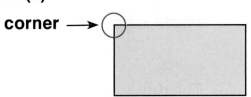

corner →

count

You **count** to find the number.

cube

cup

cylinder

data

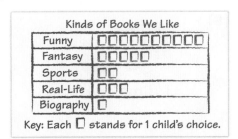

date

November						
Sunday	Monday	Tuesday	Wednesday	Thursday	Friday	Saturday
	1	2	3	4	5	6
7	8	9	10	11	12	13
14	15	16	17	18	19	20
21	22	23	24	25	26	27
28	29	30				

The **date** is November 28.

day

The **day** is Monday.

November						
Sunday	Monday	Tuesday	Wednesday	Thursday	Friday	Saturday
	1	2	3	4	5	6
7	8	9	10	11	12	13
14	15	16	17	18	19	20
21	22	23	24	25	26	27
28	29	30				

decimal point

$1.00

decimal point

difference

$$6 - 4 = 2$$

The **difference** is 2.

digit

51

tens **digit** ⟶ ↑ ↑ ⟵ ones **digit**

51 has two **digits.**

dime(s)

1 **dime** = 10¢

division, divide

3 sets of 2

division fact

$$10 \div 5 = 2$$

$$4\overline{)12} \quad 3$$

dollar(s)

$1.00

dollar sign

$1.00

↑ **dollar sign**

double

Both addends are the same in a **doubles** fact.

$$4 + 4 = 8$$

east

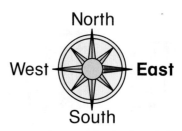

North

West — East

South

edge

edge ⟶

356

equal to

25 is **equal to** 25.

equally likely

Yellow and blue are **equally likely** to be pulled from the bag.

equivalent

The sets are **equivalent.**

estimate

To find about how many or how much

even

0, 2, 4, 6, 8, 10

exact

The **exact** time is 1:57.

face

face ⟶

fact family

6, 7, and 13 are the numbers in this **fact family.**

$$6 + 7 = 13 \qquad 7 + 6 = 13$$
$$13 - 6 = 7 \qquad 13 - 7 = 6$$

factor

$$5 \times 3 = 15$$

factors

Fahrenheit

fair share

feet

More than one foot

fewest

Ways to Make 24¢

D	P	Number of Coins
0	24	24
1	14	15
2	4	6

2 dimes and 4 pennies are the **fewest** coins that can be used to make 24¢.

fifths

flip

foot

A sheet of notebook paper is about 1 **foot** long.

1 **foot** = 12 inches

fourths

fraction

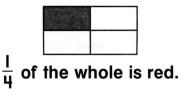

$\frac{1}{4}$ of the whole is red.

gallon

1 **gallon** = 4 quarts

gram

about 1 **gram**

graph

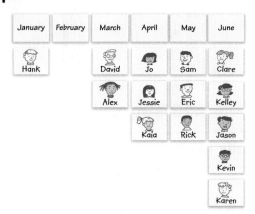

greater

Has more

greater than (>)

63 is **greater than** 29.
63 > 29

grid

grow

The pattern **grows.**

half

Half of the cubes are red.

half dollar

1 **half dollar** = 50¢

half hour

There are 30 minutes in a **half hour.**

halves

heavier, heaviest

The cherry is the **heaviest** object.
The cherry is **heavier** than the feather.

hexagon

horizontal

There are 2 **horizontal** lines.

hour(s)

There are 60 minutes in 1 **hour.**

hundreds

2 **hundreds** = 200

impossible

It is **impossible** to spin blue.

inch

intersection

jump

1 + 3 = 4

There are three **jumps** between
1 and 4.

key

key

kilogram

about 1 **kilogram**

labels

landing number

$1 + 3 = 4$

The **landing number** is 4.

landmark

The **landmark** time is 8:30.

least

The glass holds the **least.**

length

The spoon has a **length** of 6 inches.

less

The milk box holds **less** than the pot.

less likely

You are **less likely** to pick yellow than red.

less than (<)

29 is **less than** 63.
$29 < 63$

level

The white cube is at **level** B.

lighter, lightest

A pencil is **lighter** than an apple.
The pencil is the **lightest** object.

likely

Blue is a **likely** outcome.

line

line of symmetry

line of symmetry ↑

list

← **list**

liter

1 **liter** = 1,000 milliliters

long

A pen is about 6 inches **long.**

measure

measurement

The **measurement** for the side
is 1 inch.

mental math

Solving math problems without paper,
a pencil, or a calculator

meter

Unit used to measure distances and
the length of longer objects

middle

middle →

milliliter

1 **milliliter**

minute(s)

I minute

mirror image, mirror line

Both parts match.

mirror line

missing factor

$$9 \times ? = 18$$

missing factor

model

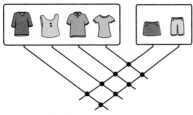

A **model** of multiplication

month

January

Sunday	Monday	Tuesday	Wednesday	Thursday	Friday	Saturday
				1	2	3
4	5	6	7	8	9	10
11	12	13	14	15	16	17
18	19	20	21	22	23	24
25	26	27	28	29	30	31

more

The pot holds **more** than the milk box.

more likely

You are **more likely** to pick green than blue.

more than

6 is **more than** 4.

most

The jar with the red lid has the **most** coins.

multiple

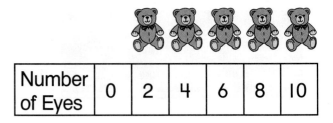

Number of Eyes	0	2	4	6	8	10

multiple choice

68
+47

Ⓐ 115
Ⓑ 105
Ⓒ 100
Ⓓ 21

multiplication, multiply

5 sets of 3

multiplication fact

$$5 \times 3 = 15 \qquad \begin{array}{r} 2 \\ \times 6 \\ \hline 12 \end{array}$$

multiplication table

nearest

20 21 22 23 24 25 26 27 28 29 30

The **nearest** ten to 26 is 30.

never

Blue can **never** be picked.

nickel(s)

1 **nickel** = 5¢

noon

Noon

12:00 P.M.

north

number line

0 1 2 3 4 5 6

number pair

2 + 8

5 + 5

7 + 3

Each **number pair** has a sum of ten.

number sentence

$$4 + 2 = 6$$
$$9 - 2 = 7$$
$$5 \times 2 = 10$$

o'clock

The clock shows 10 **o'clock.**

odd

1, 3, 5, 7, 9

one dollar

$1.00

one fourth

one half

one third

one twelfth

ones

2 **ones** = 2

open figures

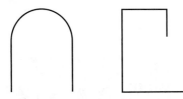

order

0   30   58   61   85   100   302

These numbers are in **order** from the smallest to the biggest.

organized list

ounce

A slice of bread weighs about 1 **ounce**.

path

pattern(s)

30, 40, 50, 60, 70 . . .

pattern unit

pattern unit

penny, pennies

I **penny** = I ¢

pentagon

perimeter

The distance around a figure

2 centimeters

2 centimeters 2 centimeters

2 centimeters

The **perimeter** measures
8 centimeters.

pictograph

Hand We Use to Write			
right	☺	☺	☺
left	☺	☺	

Key: Each ☺ stands for 5 children.

picture graph

Sports We Play				
tennis	⚪	⚪	⚪	
football	🏈	🏈		
basketball	🏀	🏀	🏀	🏀

pint

I **pint** = 2 cups

polygon

Each figure is a **polygon.**

possible

It is **possible** to pick a blue marble.

pound

A loaf of bread weighs
about I **pound.**

predict, prediction

I **predict** the next block will be blue.

prism

product

4 × 5 = 20

product

pyramid

quadrilateral

quart

I **quart** = 4 cups

quarter

I **quarter** = 25¢   **quarter** after 10

real-object graph

reasonable

99 × 6

600 would be a **reasonable** estimate.

rectangle

rectangular prism

reflection

A **reflection** looks like the figure has been flipped.

regroup

repackage

repeat

This pattern **repeats.**

rounding

Estimating to the nearest ten

23 is closer to 20 than to 30.

row(s)

row →

ruler

inch ruler

centimeter ruler

rule(s)

Add 2	
4	6
6	8
8	10

The **rule** is add 2.

same

 is the **same** as

second

separate

You can **separate** the blue cubes from the red cubes.

set(s)

3 **sets** of 4

shortest path

The **shortest path** from A to B is WWNN or NNWW.

side(s)

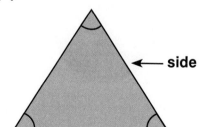

This figure has 3 **sides.**

similar

sixths

skip-count(ing)

460, 462, 464, 466, 468,...

smallest

115   15   5

5 is the **smallest** number.

solution

The way to solve a problem

south

sphere

square

square unit

12 square units

starting number

1 + 3 = 4

The **starting number** is 1.

subtract

To take away objects from a group or to compare groups

subtraction

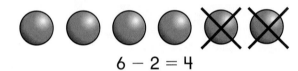

6 − 2 = 4

sum

6 + 3 = 9

The **sum** is 9.

symbol

symmetry

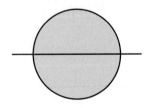

table

Our Favorite Places	
Place	**Tally**
arcade	ⵀ ⵀ
park	ⵀ
beach	ⵀ ⵀ ⵀ

tall

This tower is four cubes **tall.**

tally

1	2	3	4	5

teens

The numbers 13, 14, 15, 16, 17, 18, and 19

temperature

The **temperature** is 40°F.

ten(s)

2 **tens** = 20

thermometer

thirds

time

The **time** is 3:00.

today

The day that is right now

tomorrow

The day that is after today

ton

A whale would be weighed in **tons.**

top

total

There is a **total** of 127.

total cost

total cost = $4.50

total value

total value = 50¢

towers

trade

translate

The symbols **translate** to 124.

triangle

twelfths

uncertain

Blue is an **uncertain** outcome.

units

The pen is about 4 **units** long.

unlikely

It is **unlikely** the spinner
will land on yellow.

value

693

The **value** of the blue digit is 90.

vertex, vertices

vertex ⟶

vertical

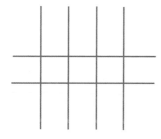

There are 4 **vertical** lines.

week

There are 7 days in 1 **week**.

weight

west

North

West⟶East

South

whole number

A number without a fraction
1, 2, 3, 4, 5, 6, 7, 8, 9 . . .

wide

The cube is about 1 inch **wide.**

word problem

If Jack jumped 5 times, and Jill
jumped 10 times, how many times
did they jump altogether?

yard(s)

1 **yard** = 3 feet

year

There are 12 months in 1 **year**.

yesterday

The day before today